XV International Workshop on Low Energy Positron and Positronium Physics 2009

Satellite Workshop of the XXVI ICPEAC 2009
Journal of Physics: Conference Series Volume 199

Toronto, Ontario, Canada
29 July-1 August 2009

Editors:

R. I. Campeanu
J. W. Darewych

A. D. Stauffer

ISBN: 978-1-61738-278-9

Printed from e-media with permission by:

Curran Associates, Inc.
57 Morehouse Lane
Red Hook, NY 12571

Some format issues inherent in the e-media version may also appear in this print version.

Printed by Curran Associates, Inc. (2010)

Institute of Physics
Dirac House, Temple Back
Bristol BS1 6BE UK

Phone: 44 1 17 929 7481
Fax: 44 1 17 920 0979

techtracking@iop.org

Additional copies of this publication are available from:

Curran Associates, Inc.
57 Morehouse Lane
Red Hook, NY 12571 USA
Phone: 845-758-0400
Fax: 845-758-2634
Email: curran@proceedings.com
Web: www.proceedings.com

TABLE OF CONTENTS

POSITRONIUM AND POSITONIUM NEGATIVE ION

Fragmentation in Positronium Collisions with Atoms ... 1
S. J. Brawley, G. Laricchia

Probing the Energy Structure of Positronium with a 203 GHz Fabry-Perot Cavity ... 9
T. Suehara, A. Miyazaki, A. Ishida, T. Namba, S. Asai, T. Kobayashi, H. Saito, M. Yoshida, T. Idehara, I. Ogawa, S. Kobayashi, Y. Urushizaki, S. Sabchevski

Ps- Emission from Cs Coated Surfaces ... 15
K. Michishio, T. Tachibana, H. Terabe, A. Miyamoto, Y. Nagashima

Precise Measurement of HFS of Positronium ... 20
A. Ishida, G. Akimoto, K. Kato, T. Suehara, T. Namba, S. Asai, T. Kobayashi, H. Saito, M. Yoshida, K. Tanaka, A. Yamamoto, I. Ogawa, S. Kobayashi, T. Idehara

Magnetised Positronium ... 25
D. P. van der Werf, C. J. Baker, D. C. S. Beddows, P. R. Watkeys, C. A. Isaac, S. J. Kerrigan, M. Charlton, H. H. Telle

Positronium Formation Cross-sections for Xe, CO_2 and N_2 ... 31
D. A. Cooke, D. J. Murtagh, G. Laricchia

ANTIHYDROGEN

Detailed Calculations on Low-energy Positron-hydrogen-molecule and Helium-antihydrogen Scattering ... 37
E. A. G. Armour, J. N. Cooper, M. R. Gregory, S. Jonsell, M. Plummer, A. C. Todd

Simulations of Antihydrogen Formation in a Nested Penning Trap ... 50
S. Jonsell, D. P. van der Werf, M. Charlton

AEGIS: An Experiment to Measure the Gravitational Interaction Between Matter and Antimatter ... 56
Michael Doser

Leading-order Relativistic Corrections to the Zeeman Splitting of Hyperfine Structure Levels in Two-fermion Bound-state Systems ... 63
Andrei G. Terekidi, Jurij W. Darewych, Marko Horbatsch

ANNIHILATION, BOUND AND RESONANT STATES

Excited, Bound and Resonant Positron-atom Systems ... 68
M. W. J. Bromley, J. Mitroy

Analysis of Experimental Positron-molecule Binding Energies ... 73
J. R. Danielson, J. A. Young, C. M. Surko

Resonant Positron Annihilation in Ammonia ... 79
G. F. Gribakin

Relaxed Geometries and Dipole Moments of Positron Complexes with Diatomic Molecules ... 85
Denise Assafrão, José R. Mohallem

PLASMAS AND SWARMS

Elastic Scattering of Slow Electrons by Positronium Atoms in Weakly-coupled Plasmas ... 91
Arijit Ghoshal, Sabyasachi Kar, Y. K. Ho

On New Developments in the Physics of Positron Swarms ... 95
Z. Lj Petrović, A. Banković, S. Dujko, S. Marjanović, M. Šuvakov, G. Malović, J. P. Marler, S. J. Buckman, R. D White, R. E. Robson

POSITRON SCATTERING BY ATOMS AND MOLECULES

Theoretical Calculation of Total Cross Sections for e^+ - NH_3 Molecule at Low Energies 101
Arvind Kumar Jain, Indira Dubey, Harsh Mohan

Ionization of the Water Molecule by Electron and Positron Impact 105
I. Tóth, R. I. Campeanu, V. Chiş, L. Nagy

Energy Scale Determination and Energy Resolution in Positron Total Cross Section Measurements 110
G. P. Karwasz, R. S. Brusa, D. Pliszka

Differential Ionization Studies by Positron Impact 115
Á. Kövér, D. J. Murtagh, A. I. Williams, G. Laricchia

Convergence Study of the Close-coupling Approach to Positron-helium Collisions 120
R. Utamuratov, A. S. Kadyrov, D. Fursa, I. Bray

Classical Trajectory Monte Carlo Calculation of the Fully Differential Cross Section for Ionization of H_2 by Positron Impact 124
R. O. Barrachina, J. Fiol

Low Energy Elastic Scattering of Positrons by Argon Atoms 129
Harsh Mohan, Arvind Kumar Jain, Sunita Sharma

An Electrostatic Slow Positron Beam Apparatus for Positron Scattering Experiments 134
Kazuaki Nagumo, Yuna Nitta, Daiki Tomida, Masamitsu Hoshino, Hiroshi Tanaka, Y. Nagashima

Progress Towards a Positron Reaction Microscope 138
A. I. Williams, Á. Kövér, D. J. Murtagh, G. Laricchia

Author Index

This volume of *Journal of Physics: Conference Series* contains the Proceedings of the XV International Workshop on Low Energy Positron and Positronium Physics, which was held at York University, Toronto, Canada, between 29 July and 1 August 2009. This Workshop was one of the official satellites of the XXVI International Conference on Photonic, Electronic and Atomic Collisions, held at Kalamazoo, MI, USA between 22 and 28 July 2009. Our Workshop was run in parallel with the XVI International Conference on Electron-Molecule Collisions and Swarms, enabling delegates to select talks from either meeting. The proceedings of the XVI International Conference on Electron-Molecule Collisions and Swarms are being published in a separate volume of *Journal of Physics: Conference Series*.

The combined attendance at the two meetings was 110, from 19 countries, with interest being roughly split between the two areas. There were 5 plenary lectures addressed to all participants and on the Positron Workshop side there were 20 invited talks and 40 posters presented in two poster sessions. The subjects covered antihydrogen, positron collisions with atoms and molecules, annihilation, bound and resonant states, positronium and positronium negative ion, plasmas and swarms. The talks and posters highlighted the vigour of the field and underlined the impressive new developments now in progress and which promise to transform the subject.

The next Workshop of this biennial series of meetings will be held in July 2011 at Maynooth University in Ireland.

R I Campeanu
J W Darewych
A D Stauffer
Guest Editors

INTERNATIONAL ADVISORY COMMITTEE FOR POSITRON WORKSHOP

Roberto S Brusa (Italy)
Michael Charlton (UK)
Arnab S Ghosh (India)
Franco A Gianturco (Italy)
Gleb F Gribakin (UK)
John Humberston (UK)
Helge Knudsen (Denmark)
Akos Kover (Hungary)
Gaetana Laricchia (UK)
Marco AP Lima (Brazil)
Allen Mills (USA)
Yasuyuki Nagashima (Japan)
Clifford M Surko (USA)
James Walters (UK)
Sandra Ward (USA)

LOCAL ORGANIZING COMMITTEE FOR POSITRON WORKSHOP

Radu I Campeanu (York, Chairman)
Jurij W Darewych (York)
Allan D Stauffer (York)
Cody Storry (York)

All papers published in this volume of *Journal of Physics: Conference Series* have been peer reviewed through processes administered by the proceedings Editors. Reviews were conducted by expert referees to the professional and scientific standards expected of a proceedings journal published by IOP Publishing.

Fragmentation in positronium collisions with atoms

S J Brawley and G Laricchia

UCL Department of Physics and Astronomy, University College London, Gower Street, London, WC1E 6BT, UK

Abstract. Recent positronium collision experiments are reviewed in which the cross-sections for projectile- and/or target-fragmentation processes are determined. In contrast with theory, the likelihood for the latter process appears to be significant for positronium collisions with Xe at only ~3 eV above threshold. Measurements of the total direct ionization cross-section by positron impact, performed to test the reliability of the experimental apparatus and method, are also presented .

1. Introduction

Positronium (Ps) is the bound state of an electron, e^-, and its antimatter partner, the positron, e^+. It has half the reduced mass of hydrogen and thus its principal energy levels are correspondingly halved. Over the past twenty years, its controlled production has enabled investigations ranging from its reflection from LiF [1] to its fragmentation in collisions with He [2] and Xe [2, 3].

Fast positronium may be produced by neutralizing a positron beam, for example, by transmission through a thin foil [4], a gas [5, 6, 7, 8, 9], or by scattering it from a surface [10, 11]. The relative merits of these methods with respect to efficiency of production, energy and angular distributions, as well as quantum state population have been previously appraised [12]. At UCL, a beam of Ps atoms is produced using the gas technique according to the charge-exchange reaction [9, 13]:

$$e^+ + A \longrightarrow Ps + A^+. \tag{1}$$

Several gas targets (A) have been tried, the most efficient being H_2 at low energies [14] and N_2 up to 250 eV [15]. For these targets, the energy of the Ps beam, E_{Ps}, has been found to be consistent with the formation of ground-state Ps according to:

$$E_{Ps} = E_+ - E_i + \frac{6.8\,eV}{n^2}, \tag{2}$$

where E_+ is the energy of the positron beam, E_i is the first ionisation energy of the production gas and $6.8\,eV/n^2$ is the binding energy of the Ps atoms of principal quantum number n. The formation of Ps in an excited-state has recently been found to make an increasingly significant contribution — of maximum probabilities 0.06, 0.12 and 0.23, respectively — to the total Ps formation cross-section of

Table 1. Lowest energy reactions in Ps - He scattering in which a free positron and/or electron is produced in the final state.

	Reaction	Threshold (eV)	Description
(a)	$Ps + He \longrightarrow e^+ + e^- + He$	6.8	Ps fragmentation
(b)	$Ps + He \longrightarrow Ps^- + He^+$	24.2	Ps negative ion formation
(c)	$Ps + He \longrightarrow Ps + e^- + He^+$	24.5	Target-ionisation
(d)	$Ps + He \longrightarrow e^+ + e^- + He*$	27.0	Ps fragmentation and target-excitation
(e)	$Ps + He \longrightarrow Ps* + e^- + He^+$	29.7	Ps excitation and target-ionisation
(f)	$Ps + He \longrightarrow e^+ + 2e^- + He^+$	31.4	Ps fragmentation and target-ionisation

He, Ar and Xe [16]. However, the collimated-Ps yields from the latter target have been found to be dominated by formation in the $n = 1$ state and, at the same time, surprisingly low [13] when considering the large integrated Ps formation cross-section for this target e.g. [17]. This may be the result of a broad differential cross-sections for Ps($n = 1, 2$) formation and/or of Ps being formed in an excited state and lost from the beam within the production cell itself due to a larger total cross-section. As previously noted in [12], since fast Ps atoms are formed from solid neutralizers by epithermal positrons, they have a broad energy spread and may be distributed over several quantum states e.g. [18, 19]; so that, depending on experimental conditions, the contribution of excited-state Ps to a beam could be significant, especially for S-states whose lifetimes scale approximately as n^3.

Over the past two decades, the Ps beam at UCL has been used to measure total cross-sections, Q_T^{Ps} [14, 15, 20, 21, 22, 23], and more recently fragmentation cross-sections [2, 3, 22]. In collisions with atoms and molecules, Ps may break-up if its kinetic energy exceeds 6.8 eV and experimentally this may be probed by detecting the residual positron or electron. Whilst detection of the positron is a unique signature of Ps fragmentation, detection of the electron may also indicate target ionisation.

2. Fragmentation in Ps–He collisions

Experimental investigations were carried out on He by Armitage et al. [2] in which the integrated fragmentation cross-sections, Q_f^+, were determined for the ejected positron at Ps energies $E_{Ps} = 13$, 18, 25 and 33 eV as shown in Figure 1. Detection of the ejected positrons encompasses reactions (a), (d) and (f) in Table 1. In the study of Armitage et al. [2], the longitudinal energy distributions of the ejected positrons were also determined via a time-of-flight technique. As shown in Figure 2, a peak was found just below half of the residual energy, $E_r = E_{Ps} - 6.8$ eV, indicative of the electron loss to the continuum (ELC) process in which the ionised projectile and the ejected electron remain strongly correlated with a small relative velocity. The shift of the peak to lower energies from $E_r/2$ observed was ascribed to the ejected positrons being emitted in the laboratory frame within a ~20° cone around the beam axis; a result supported by a classical trajectory Monte Carlo (CTMC) [24] and an impulse approximation calculation [25], the former agreeing in shape and the latter in shape and magnitude with the experimental data. In addition to the errors arising from statistical uncertainties in the absolute scale in Figures 1 and 2, there is a systematic uncertainty of (+8, -(20-30))% arising from the determination of detection efficiencies. It is an energy-independent error which equally affects each value of Q_f^+ and dQ_f^+/dE_ℓ .

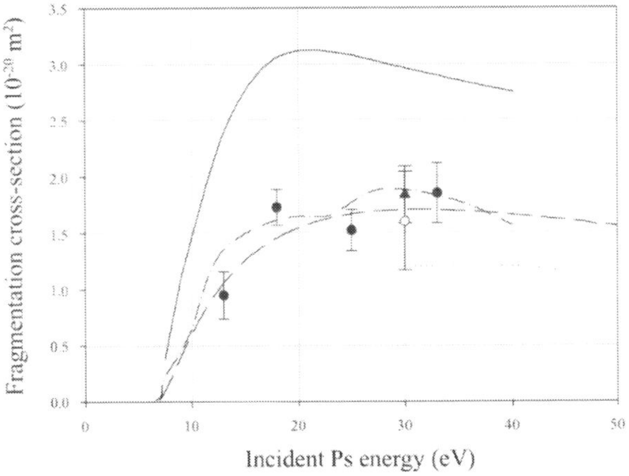

Figure 1. Integrated cross-sections for He. Experiment: Q_f^+: •, [2]; ▲, [22]. Q_f^-: ○, [22]. Theory: ———, CTMC [24]; — · — · —, coupled-state calculation [26]; — — —, impulse approximation [25]; ·········, Coulomb–Born approximation [27]. An additional uncertainty of (+8, -(20-30))% exists on the absolute scale for the experimental Q_f^+ and Q_f^-.

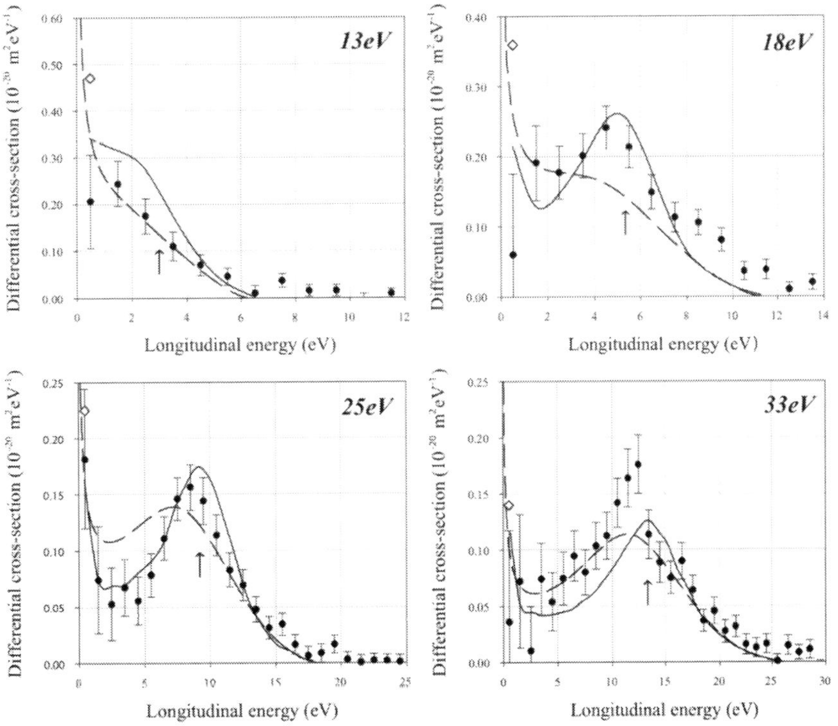

Figure 2. Differential fragmentation cross-sections, dQ_f^+/dE_ℓ, for Ps+He collisions at 13, 18, 25 and 33eV Ps incident energy. •, [2]; ———, CTMC [24] multiplied by 0.5; — — —, impulse approximation [25]. Average $dQ_f^+/dE_\ell < 1$ eV: ◇, [25]. ↑ indicates the position of $E_r/2 = (E_{Ps} - 6.8\,\text{eV})/2$.

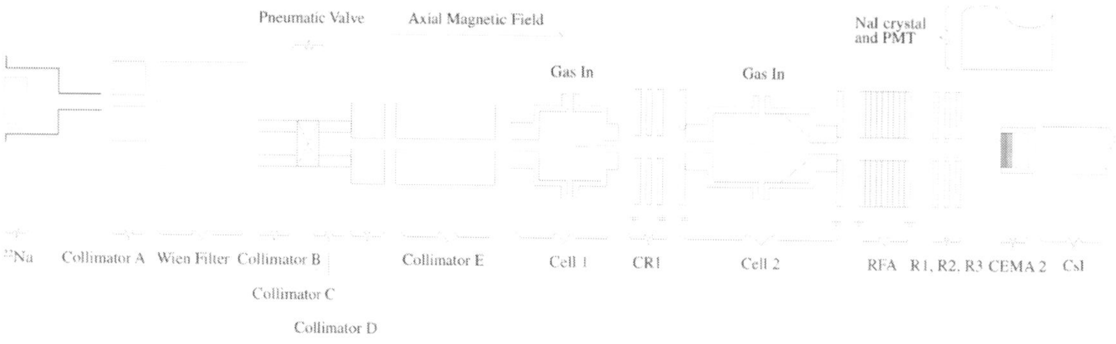

Figure 3. A schematic of the positronium beam line at UCL. Production of the near-monoenergetic positron beam is discussed in detail elsewhere, e.g. [15].

When detecting the ejected electrons from Ps–He collisions, the experiment is sensitive to all the reactions in Table 1. Thus the almost equality between Q_{f}^{-} and Q_{f}^{+}, measured by Armitage et al. [22], implies a negligible cross-section for target-ionisation, $Q_{\mathrm{i}}^{\mathrm{Ps}}$, at $E_{\mathrm{Ps}} = 30\,\mathrm{eV}$ (see Figure 1). This result is supported by the First Born approximation calculation of Walters et al. [28] employed to determine the contribution from target-inelastic effects — i.e. target-excitation and -ionisation — to supplement the earlier, target-elastic impulse approximation of Starrett et al. [25]. It is noted that should the Ps negative ion, Ps^{-}, be produced, its lifetime of 0.48 ns against annihilation [29] means an electron would remain available for detection and contribute to the measured Q_{f}^{-}.

To determine Q_{f}^{-}, the Ps beam, produced in cell 1 via the charge exchange reaction of Equation 1, was impinged against the target gas in cell 2 (see Figure 3). Any positrons not neutralised were reflected by a positive potential on CR1. Cell 2 was floated at a negative potential, V_{cell}, to accelerate the ejected electrons (from Ps fragmentation or target-ionisation) to energies much greater than secondary electrons released from surfaces. In this way, electrons from Ps–He collisions were separated from the background, and the longitudinal energies of the signal-electrons were analysed by a pair of retarding grids, R2 and R3. Electrons and positrons ejected from cell 2 were confined by an axial magnetic field of 22 mT and were detected by a channel electron multiplier array (CEMA). Whilst positrons ejected in the backwards direction were reflected by CR1 and thus detected, electrons ejected backward were lost due to attenuation. For this reason, all present experimental values of Q_{f}^{-} should be considered to be lower limit estimates.

3. Fragmentation in Ps–Xe collisions

Preliminary investigations for Xe were performed by Armitage et al. [22], in which Q_{f}^{-} — determined using the retarding field method above — was found to be almost a factor-of-two larger than Q_{f}^{+} at $E_{\mathrm{Ps}} = 30\,\mathrm{eV}$, suggesting the occurrence of target-ionisation (see Table 2). This work was supplemented by that of Brawley et al. [3], as shown in Figure 4, where the results of $dQ_{\mathrm{f}}^{+}/dE_{\ell}$ for 18 and 30 eV Ps collisions are compared to an available theory [28]. Here $dQ_{\mathrm{f}}^{+}/dE_{\ell}$ may be seen to display a peak just below $E_{\mathrm{r}}/2$ — consistent with the ELC process, as observed in the time-of-flight work for He [2]. However, it should be noted that the extraction efficiency of positrons with energies below 1 eV from the main body of the cell is estimated to be $\lesssim 50\%$. Figure 4 also shows the results of $dQ_{\mathrm{f}}^{+}/dE_{\ell}$ at $E_{\mathrm{Ps}} = 30\,\mathrm{eV}$ and those of the target-elastic impulse approximation theories of Walters et al. [28] at $E_{\mathrm{Ps}} = 33\,\mathrm{eV}$. At this Ps impact energy, reactions (d) and (f) in Table 2 are energetically accessible, in

Table 2. Lowest energy reactions in Ps - Xe scattering in which a free positron and/or electron is produced in the final state.

	Reaction	Threshold (eV)	Description
(a)	$Ps + Xe \longrightarrow e^+ + e^- + Xe$	6.8	Ps fragmentation
(b)	$Ps + Xe \longrightarrow Ps^- + Xe^+$	11.8	Ps negative ion formation
(c)	$Ps + Xe \longrightarrow Ps + e^- + Xe^+$	12.1	Target-ionisation
(d)	$Ps + Xe \longrightarrow e^+ + e^- + Xe^*$	15.1	Ps fragmentation and target-excitation
(e)	$Ps + Xe \longrightarrow Ps^* + e^- + Xe^+$	17.2	Ps excitation and target-ionisation
(f)	$Ps + Xe \longrightarrow e^+ + 2e^- + Xe^+$	18.9	Ps fragmentation and target-ionisation

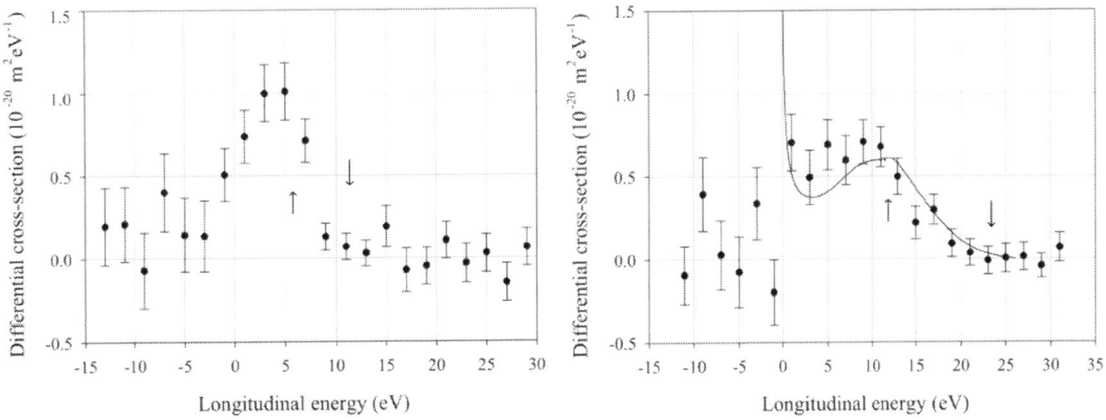

Figure 4. dQ_f^+/dE_ℓ for Ps collisions with Xe at 18 eV and 30eV. •, [3] at E_{Ps} = 30 eV; ———, [28] for E_{Ps} = 33 eV; ↑, $E_r/2$; ↓, E_r.

addition to reaction (a), and may account for the differences between these results and the target-elastic data [28].

Measurements of Q_f^+ and Q_f^- by Brawley et al. [3] are shown in Figure 5 together with the calculations of Walters et al. [28] and Starrett et al. [30]. An additional uncertainty in the absolute scale of (+30, -50)% of the same kind as described above applies to these measurements. As before, due to the fact that backward-ejected electrons remain undetected, Q_f^- must be considered a lower limit. Also, the difference between Q_f^- and Q_f^+ yields Q_i^{Ps}, for processes in which target-ionisation occurs without Ps fragmentation, namely reactions (b), (c) and (e) in Table 2. However, some overestimate from reaction (f) may occur if the two electrons in the final state are separated sufficiently at the CEMA to be detected independently. Considering the variety of energies and ejection angles that the two electrons may have, it seems likely that this will occur. Once again, the experimental Q_f^- may include Ps^- formation not yet included in the theories.

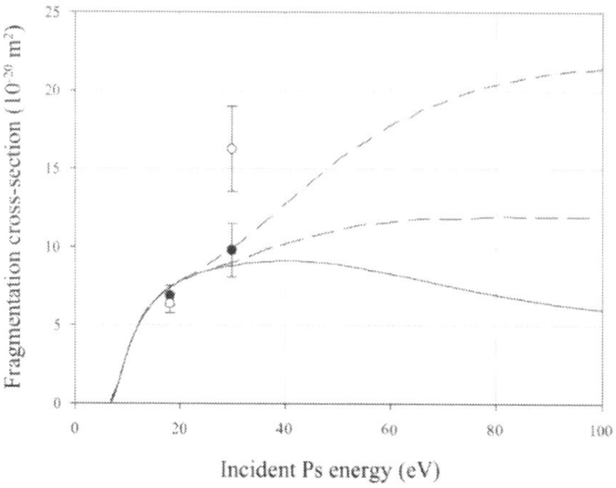

Figure 5. Q_f^+ and Q_f^- for Ps collisions on Xe. Experiment: •, Q_f^+; ○, Q_f^-. Theory: ———, target-elastic integrated fragmentation work of [28] for the ejection of positrons; — — —, Q_f^+ work of [30]; — · — · —, Q_f^- work of [30]. There is an additional uncertainty of (+30, -50)% on the absolute scale for the experimental Q_f^+ and Q_f^-.

In view of the excess of Q_f^- over Q_f^+ and the considerable difficulties in discriminating signal electrons from the significant backgrounds of secondary electrons, further tests were undertaken to gauge the reliability of the experimental system and method. These consisted of measurements of the direct ionisation cross-section for positron impact (Q_i^+). This has been determined at 42 and 62 eV for Ar, and at 40 and 55 eV for Xe. As in the case of Ps impact, ejected electrons are confined by the magnetic field — of the same strength as for the Ps experiments — and accelerated by V_{cell} to be detected by the CEMA plates (see Figure 3). For this experiment, the positron collision energy in the cell is $E_+^{cell} = E_+ + eV_{cell}$, and thus a value of V_{cell} = -22 V was chosen in order to accelerate the ejected electrons whilst not exceeding the maximum transverse energy of 64 eV which may be confined by the field. A positive potential of 250 V is applied to an electrode (RFA) to reflect any positrons that emerge from the scattering cell. The emerging electrons that reach the detection region are energy analyzed in the same way as in the case of the Ps fragmentation experiments above. The background originating from secondary electrons is taken into account and subtracted in order to determine the net number of electrons from ionisation events.

The ionisation cross-section by positron impact, Q_i^+, was then calculated as follows:

$$Q_i^+ = \frac{N_-}{N_{scatt}^+} Q_T^+ \left(\frac{\epsilon_+}{\epsilon_-} \right), \tag{3}$$

where N_- is the net number of electrons detected from positron collisions, N_{scatt}^+ is the number of scattered positrons in cell 2, Q_T^+ is the positron total cross-section for the target gas, and the last term is the ratio of the detection efficiency of positrons to that of electrons by the CEMA plates (in this work equal to 1.60 ± 0.06 [31]). Whilst only electrons emitted within ± 90° of the central beam axis may be detected, most of the contribution (~90%) to N_- from the backward angles are included in the measurements by reflecting the transmitted positrons at the RFA. Although multiple-ionisation accounts for ≤ 10% of all direct ionisation events [32], it may contribute to the present cross-sections and, in principle, with enhanced sensitivity due to the availability of more than one electron in the final state. Thus in comparing with

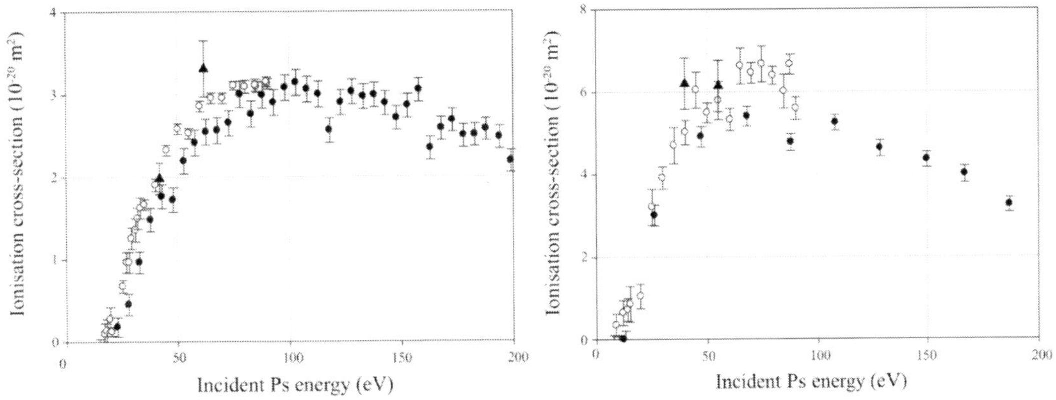

Figure 6. LHS: Ionisation cross-sections for positron impact on Ar. Q_i^{n+}: ▲, present results; ○, [33]. Q_i^+: ●, [34]. RHS: Ionisation cross-sections for positron impact on Xe. Q_i^{n+}: ▲, present results; ○, [33]. Q_i^+: ●, [35]. An additional uncertainty of ± 5% applies to the absolute scale of the present measurements.

other measurements, it is more meaningful to label the present measurements as Q_i^{n+}, equal to the sum of the cross-sections for all degrees of direct ionisation.

The present absolute results of Q_i^{n+} for Ar and Xe — as calculated using Equation 3 — are shown in Figure 6. In both cases, there is an additional error in the absolute scale of ± 5% in the present data due to uncertainties in calculating (ϵ_+/ϵ_-). For Ar, there is consistency between the present Q_i^{n+} and those of Marler et al. [33] and the Q_i^+ work of Moxom et al. [34] for 42 eV positron collisions. The agreement is less pronounced at the highest energy investigated, although it is within errors. For Xe, there is a fairly good consistency at both energies between the present work and that of Marler et al. [33]. At both energies, the present Q_i^{n+} are ~10% higher than the Q_i^+ reported by Kara et al. [35].

The consistency of the present results for Q_i^{n+} with previous measurements (by two different methods), implies that contributions from secondary electrons to the background have been properly accounted for, thus affording confidence in Q_f^-.

4. Conclusion and outlook

In conclusion, Ps fragmentation and target-ionisation investigations carried out at 18 and 30 eV Ps collisions with He and Xe, suggest significant target-ionisation effects from the latter target at the higher energy, in disagreement with theory. The system reliability has been tested by performing positron impact ionization measurements which are found to be consistent with available data.

Ps fragmentation was first extensively investigated theoretically by Ludlow and Walters [36] at high impact energy (500 eV). Whilst these conditions have not yet been amenable to experimental investigations, recently, Ps⁻ (first observed by Mills [37]) has been produced with significant efficiencies [38] using a surface doping technique as suggested in [12, 32]. These developments open up the prospect of being able to produce monoenergetic Ps beams of useful intensities at high energies by accelerating Ps⁻ and by photodetaching one electron — a method first proposed by Mills and Crane [4]. Much future activity is envisaged.

Acknowledgments

We wish to thank our co-workers at UCL for useful discussions and assistance. The Science and Engineering Research Council is gratefully acknowledged for financial support.

[1] Weber M, Tang S, Berko S, Brown B, Canter K, Lynn K, Mills Jr A, Roellig L and Viescas A 1988 *Physical Review Letters* **61** 2542–2545
[2] Armitage S, Leslie D, Garner A and Laricchia G 2002 *Physical Review Letters* **89** 173402
[3] Brawley S, Beale J, Armitage S, Leslie D, Kover A and Laricchia G 2008 *Nuclear Instruments and Methods in Physics Research B* **266** 497–501
[4] Mills Jr A and Crane W 1985 *Physical Review A* **31** 593–597
[5] Brown B 1985 *Positron Annihilation: Proceedings of the Seventh International Conference on Positron Annihilation* p 328
[6] Brown B 1986 *Positron (Electron) – Gas Scattering, Proceedings of the Third International Workshop on Positron Electron-Gas Scattering* (World Scientific) pp 212–221
[7] Laricchia G, Charlton M, Griffith T and Jacobsen F 1986 *Positron (Electron) – Gas Scattering, Proceedings of the Third International Workshop on Positron Electron-Gas Scattering* (World Scientific)
[8] Laricchia G, Charlton M, Davies S, Beling C and Griffith T 1987 *Journal of Physics B: Atomic, Molecular and Optical Physics* **20** L99–105
[9] Laricchia G, Davies S, Charlton M and Griffith T 1988 *Journal of Physics E: Scientific Instruments* **21** 886–888
[10] Gidley D, Mayer R, Frieze W and Lynn K 1987 *Physical Review Letters* **58** 595—598
[11] Howell R, Rosenberg I and Fluss M 1986 *Physical Review B* **34** 3069–3075
[12] Laricchia G 1995 *Positronium beams and surfaces* (IOS, Amsterdam) chap 3, pp 401–419
[13] Laricchia G, Armitage S and Leslie D 2004 *Nuclear Instruments and Methods in Physics Research B* **221** 60–68
[14] Garner A, Laricchia G and Özen A 1996 *Journal of Physics B: Atomic, Molecular and Optical Physics* **29** 5961–5968
[15] Leslie D, Armitage S and Laricchia G 2002 *Journal of Physics B: Atomic, Molecular and Optical Physics* **35** 4819–4827
[16] Murtagh D, Cooke D and Laricchia G 2009 *Physical Review Letters* **102** 133202
[17] Laricchia G, Van Reeth P, Szluińska and Moxom J 2002 *Journal of Physics B: Atomic, Molecular and Optical Physics* **35** 2525–2540
[18] Schoepf D, Berko S, Canter K and Sferlazzo P 1992 *Physical Review A* **45** 1407–1411
[19] Day D, Charlton M and Laricchia G 2001 *Journal of Physics B: Atomic, Molecular and Optical Physics* **34** 3617–3628
[20] Zafar N, Laricchia G and Garner A 1996 *Physical Review Letters* **76** 1595–1598
[21] Garner A, Özen A and Laricchia G 1998 *Nuclear Instruments and Methods in Physics Research B* **143** 155
[22] Armitage S, Leslie D, Beale J and Laricchia G 2006 *Nuclear Instruments and Methods in Physics Research B* **247** 98–104
[23] Beale J, Armitage S and Laricchia G 2006 *Journal of Physics B: Atomic, Molecular and Optical Physics* **39** 1337–1344
[24] Sarkadi L 2003 *Physical Review A* **68** 032706
[25] Starrett C, McAlinden M and Walters H 2005 *Physical Review A: Atomic, Molecular, and Optical Physics* **72** 012508
[26] Blackwood J, Campbell C, McAlinden M and Walters H 1999 *Physical Review A* **60** 4454
[27] Ray H 2002 *Journal of Physics B: Atomic, Molecular and Optical Physics* **35** 3365–3376
[28] Walters H, Starrett C and McAlinden M 2006 *Nuclear Instruments and Methods in Physics Research B* **247** 111–116
[29] Mills Jr A 1983 *Physical Review Letters* **50** 671–674
[30] Starrett C, McAlinden M T and Walters H 2008 *Physical Review A* **77** 042505
[31] Brawley S 2009 *Collisions of positronium with atoms and molecules* Ph.D. thesis University College London
[32] Laricchia G, Armitage S, Kövér A and Murtagh D 2008 *Advances in Atomic, Molecular and Optical Physics* **56** 1–47
[33] Marler J, Sullivan J and Surko C 2005 *Physics Review A* **71** 022701
[34] Moxom J, Ashley P and Laricchia G 1996 *Canadian Journal of Physics* **74** 367
[35] Kara V, Paludan K, Moxom J, Ashley P and Laricchia G 1997 *Journal of Physics B: Atomic, Molecular and Optical Physics* **30** 3933–3949
[36] Ludlow J and Walters H 2001 *Many-Particle Spectroscopy of Atoms, Molecules, Clusters, and Surfaces* (Kluwer/Plenum)
[37] Mills Jr A 1981 *Physical Review Letters* **46** 717–720
[38] Nagashima Y, Hakodate T, Miyamoo A and Michishio K 2008 *New Journal of Physics* **10** 123029

Probing the Energy Structure of Positronium with a 203 GHz Fabry-Perot Cavity

T Suehara[1], A Miyazaki[2], A Ishida[2], T Namba[1], S Asai[2], T Kobayashi[1], H Saito[3], M Yoshida[4], T Idehara[5], I Ogawa[5], S Kobayashi[5], Y Urushizaki[5] and S Sabchevski[6]

[1] International Center for Elementary Particle Physics (ICEPP), The University of Tokyo, 7-3-1 Hongo, Bunkyo-ku, Tokyo, 113-0033, Japan

[2] Department of Physics, Graduate School of Science, The University of Tokyo, 7-3-1 Hongo, Bunkyo-ku, Tokyo, 133-0033, Japan

[3] Department of General Systems Studies, Graduate School of Arts and Sciences, The University of Tokyo, 3-8-1 Komaba, Meguro-ku, Tokyo, 153-8902, Japan

[4] Accelerator Laboratory, High Energy Accelerator Research Organization (KEK), 1-1 Oho, Tsukuba, Ibaraki, 305-0801, Japan

[5] Research Center for Development of Far-Infrared Region, University of Fukui (FIR-FU), 3-9-1 Bunkyo, Fukui, Fukui, 910-8507, Japan

[6] Bulgarian Academy of Science, 1, 15 Noemvri Str., 1040 Sofia, Bulgaria

E-mail: suehara@icepp.s.u-tokyo.ac.jp

Abstract. Positronium is an ideal system for the research of the bound state QED. The hyperfine splitting of positronium (Ps-HFS: about 203 GHz) is sensitive to new physics beyond the Standard Model via a vacuum oscillation between an ortho-Ps and a virtual photon. Previous experimental results of the Ps-HFS show $3.9\,\sigma$ (15 ppm) discrepancy from the QED calculation. All previous experiments used an indirect method with static magnetic field to cause Zeeman splitting (a few GHz) between triplet states of ortho-Ps, from which the HFS value was derived. One possible systematic error source of the indirect method is the static magnetic field. We are developing a new direct measurement system of the Ps-HFS without static magnetic field. In this measurement we use a gyrotron, a novel sub-THz light source, with a high-finesse Fabry-Perot cavity to obtain enough radiation power at 203 GHz. The present status of the optimization studies and current design of the experiment are described.

1. Introduction

Positronium (Ps), the electron-positron bound state, is a purely leptonic system. The energy difference between ortho-positronium (o-Ps, 1^3S_1 state) and para-positronium (p-Ps, 1^1S_0 state), hyperfine splitting of positronium (Ps-HFS), is a good target for precise bound state QED verification. The Ps-HFS value is approximately 203 GHz (0.84 meV), which is significantly larger than hydrogen HFS (1.4 GHz). One reason for the large Ps-HFS is the quantum oscillation: o-Ps $\rightarrow \gamma^* \rightarrow$ o-Ps (o-Ps has the same quantum number as a photon). Since some hypothetical particles (such as an axion or a millicharged particle) can participate in the quantum oscillation, resulting in a shift of Ps-HFS value, its precise measurement provides a probe for new physics beyond the Standard Model.

Measurements of the Ps-HFS have been performed in 70's and 80's with the combined precision of 3.3 ppm[1, 2]. The results were consistent with 1st order calculation of the QED available at that time. However, the 2nd and 3rd order corrections have been calculated recently, with the new prediction of 203.39169(41) GHz[3], deriving from the measured value of 203.38865(67) GHz by 3.9σ. This discrepancy may come from new physics or the common systematic errors in the previous measurements.

In all previous measurements, the Ps-HFS transition was not directly measured, since 203 GHz was too high frequency to produce and control. They measured Zeeman splitting of o-Ps instead. A static magnetic field makes Zeeman mixing between $m_z = 0$ state of o-Ps and p-Ps, the resultant energy level of $m_z = 0$ state becomes higher than $m_z = \pm 1$ (Zeeman splitting). This Zeeman splitting, which is proportional to the HFS energy level, is a few GHz frequency if ~ 1 Tesla magnetic field is applied. They applied uniform magnetic field in RF cavities where positronium was introduced, causing Zeeman transition and resultant o-Ps to p-Ps transition. However, uncertainty (especially non-uniformity) of the static magnetic field, which is very sensitive to the measured HFS value, can be a large source of systematic error.

In contrast to the indirect method, we plan to directly measure HFS transition, which does not need a static magnetic field, and is thus free from the systematic error from the field. To cause the direct transition, a powerful 203 GHz radiation field is necessary. We are developing sub-THz to THz light source called gyrotron, and also a high-finesse Fabry-Perot cavity to accumulate sub-THz photons for the direct HFS measurement.

2. Gyrotron

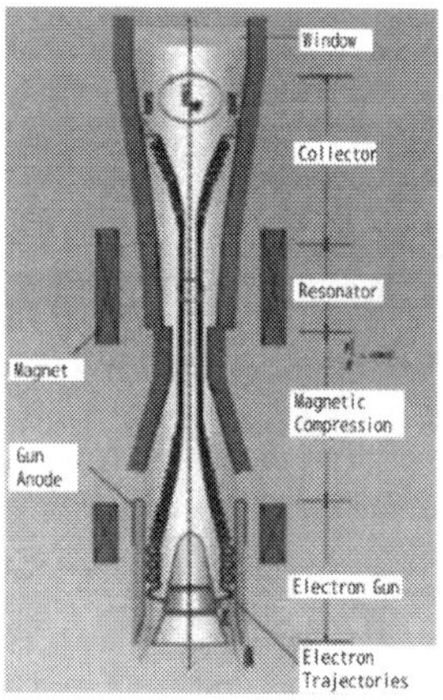

Figure 1. Schematic of gyrotrons.

Figure 2. Picture of the gyrotron for the HFS experiment.

Gyrotron[4] is a novel high power radiation source for sub-THz to THz frequency region. Structure of gyrotrons is shown in Figure 1. The electrons are produced and accelerated at the DC electron gun, concentrated and rotated as cyclotron motion in the superconducting magnet. The cyclotron frequency f_c is

$$f_c = \frac{eB}{2\pi m_0 \gamma}, \tag{1}$$

where B is the magnetic field strength, m_0 is the electron rest mass and γ is the relativistic factor of the electron. A cavity is placed at the maximum magnetic field, whose resonance frequency is tuned just to the cyclotron frequency. The electrons stimulate resonance of the cavity and produce coherent photons at the cavity. The photons are guided to the output port through the window, while electrons are dumped at the collector.

We developed a gyrotron operating at $f_c = 203$ GHz with $B = 7.425$ Tesla, $\gamma \sim 1.02$. Figure 2 shows the gyrotron. The obtained radiation power is 609 W at the window, which is reduced to 440 W during transmission through the waveguide system to the positronium cavity. The spectral width is determined by B uniformity and γ spread by thermal distribution of electrons, and is expected to be less than MHz, which is narrow enough to make resonance at the Fabry-Perot cavity. Measurement results of a similar gyrotron shows the spectral width is less than 10 kHz[5]. Frequency can be tuned by changing γ factor with different acceleration of electrons, but the tuning range is limited by resonant width of the cavity to several handreds of megahertz. Off-203 GHz radiation can be obtained by using different resonance mode of the cavity. In our gyrotron, nearest resonant frequency to 203 GHz is at around 199 GHz. The radiation power of the gyrotron will be monitored to account for corrections to the transition probability.

3. Fabry-Perot cavity

Photons produced at the gyrotron are transported and accumulated in a cavity to cause Ps-HFS transition. Since 203 GHz ($\lambda = 1.475$ mm) photons can be treated optically at the centimeter (or larger) size scale, we plan to use a Fabry-Perot cavity. Fabry-Perot cavity consists of two opposing mirrors to confine photons between them. Unlike RF cavities, the confinement in the Fabry-Perot cavity is 1-dimensional while the other four sides are open. We use a metal-mesh mirror on the input side of the cavity and use a copper concave mirror on the other side.

Table 1. Mesh parameters with simulation and measurement result. reflectance and transmittance are from estimations by the simulation, and finesse is from results of the measurements. Mesh of the last row will be used in the real experiment.

mesh material	line width	line separation	reflectance	transmittance	finesse
gold	20 μm	50 μm	99.3%	0.32%	650
gold	10 μm	50 μm	98.6%	0.75%	290
silver	50 μm	130 μm	96.9%	2.70%	180
silver	400 μm	240 μm	99.2%	0.58%	fabricating

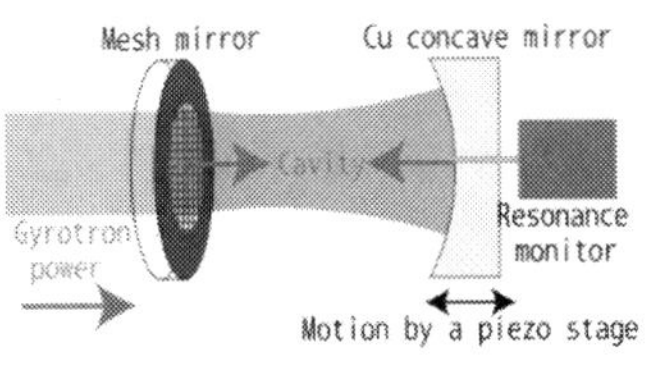

Figure 3. Overview of the cavity.

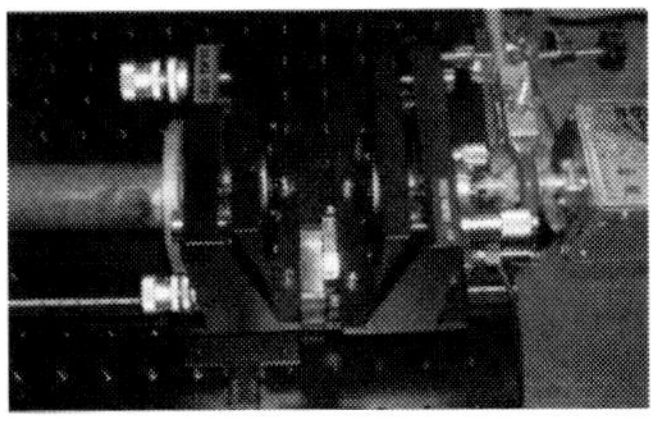

Figure 4. Picture of setup of the cavity test.

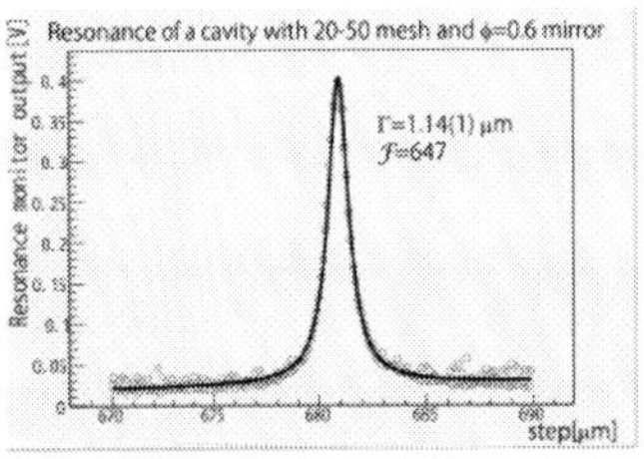

Figure 5. Resonance shape obtained by the test.

The two most important characteristics of cavities are "finesse" and "input coupling". Finesse can be written as $\mathcal{F} = 2\pi/(1-\rho)$, where ρ is the fraction of power left after one round-trip, characterizing capability of the cavity to store photons. To maximize the finesse, power losses, which are categorized into diffraction loss, medium loss and ohmic loss, must be minimized. With the confinement of photons by the concave mirror and gas medium, diffraction and medium loss are negligible in our cavity. Ohmic loss occurs at the cavity mirror, which is around 0.15% at the copper mirror and more at the mesh mirror. The ohmic loss at the mesh mirror varies by mesh parameters, which can be calculated by field simulation (see Table 1). Input coupling is the fraction of input power introduced to the cavity mode, an important parameter to introduce photons efficiently into the cavity. In our cavity, the input coupling is mainly determined by transmittance of the input mesh mirror, which is also calculated by the simulation.

Finesse and the input coupling were measured using various mesh and concave mirror parameters. In the test setup, gyrotron power was introduced to the prototype cavity, which consisted of a mesh mirror on the mirror mount and a concave mirror on a mount on a piezo stage (see Figure 3 and 4). Input, transmitted and reflected powers were monitored by three pyroelectric power monitors. By shifting cavity length precisely by the piezo stage, Breit-Wigner resonance shape was seen. Finesse could be obtained from the width of the resonance with the transmit power monitor, and the input coupling was seen by the reflection monitor. With current best combination of the mesh and the concave mirror, $\mathcal{F} = 650$ was obtained (see Figure 5). For the input coupling, concrete value could not be obtained because of interference of the reflection power and difficulty to determine absolute power of the reflection/transmission because of non-optimal setup of the power measurements. Input coupling of $\sim 20\%$ can be estimated with the reflection data.

4. Experimental Setup

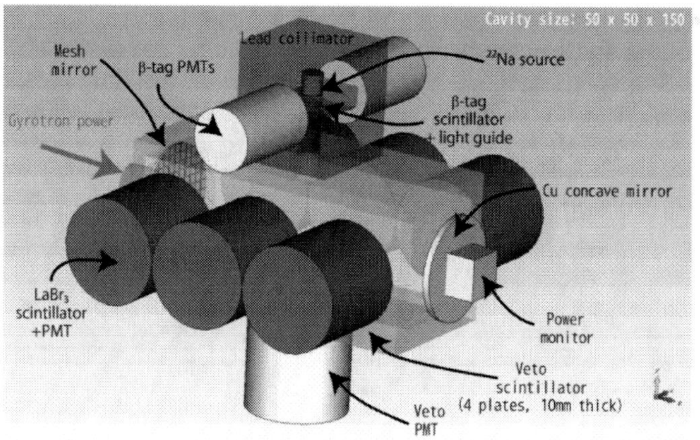

Figure 6. Overview of the experimental setup.

Figure 6 shows a schematic view of the detection system. Gyrotron power is introduced to the cavity via the mesh mirror, and accumulated between the mesh and the Cu mirror. Sodium-22 source, which emits a positron with a 1275 keV photon, is located 40 mm above the cavity. Emitted positrons pass through the β-tag scintillator to generate start timing, lead collimator (20 mm length, 15 mm aperture) and a hole on the veto scintillator to reach the cavity. Lead collimator also works as a shield to protect $LaBr_3$ scintillator from 1275 and 511 keV (annihilation) photons emitted around the source. The cavity is filled with mixed air of nitrogen and iso-C_4H_{10} to form positronium atoms (20% efficiency). p-Ps (25% of all Ps) annihilates to two 511 keV photons immediately as well as positron annihilation, while o-Ps (75%) remains with $\tau \sim 140$ ns and decays to three photons (< 511 keV), generating delayed

Table 2. Result of the Monte Carlo simulation. Accidental bg. does not include accidental tagging rate. 3 gamma decay is for the estimation of o-Ps background, and 2 gamma decay is for the estimation of signal rate. 3 gamma decay and 2 gamma decay does not include o-Ps production probability and 2 gamma decay does not include HFS transition probability. For the 511 keV selection, an energy window of 20 keV (501 to 521 keV) is used. Arrival time is not simulated. For the final estimation of signal and background rate, see the text.

trigger	all events	accidental bg. (non-tag, non-veto)	3 gamma decay inside cavity	2 gamma decay inside cavity
all	1.00×10^7	7.36×10^6	28465	28500
tagged	2.53×10^6	-	25839	25930
+non-veto	2.31×10^6	-	18714	18981
+single La	80848	1.60×10^5	7901	4967
+511 keV	14350	9751	212	2186
+double La	572	110	75	417
+511 keV both	122	3	6	114

signal at $LaBr_3$ scintillator. The six $LaBr_3$ scintillators surround the cavity to catch photons with energy resolution of $\sim$ 4% which can efficiently separate 511 keV photons (evidence of HFS transition) from photons from o-Ps decay. The $LaBr_3$ scintillators have timing resolution of $\sim$ 300 psec to separate delayed events (signal) from prompt events (annihilation or 1275 keV background).

Signal correction efficiency and background rate are estimated using Monte-Carlo simulation (GEANT4). The major background processes are o-Ps contamination and accidental photons from positron creation and annihilation without tag or veto signal which occur just around o-Ps lifetime from another event which is tagged but does not give energy deposit on $LaBr_3$ scintillators. Table 2 shows rates of the 2 gamma decay and each background with various trigger conditions. Assuming 10 kW power accumulation at the cavity (50 W introduction from gyrotron, 200 times multiplication by the cavity), HFS transition probability per o-Ps in the cavity is about 0.6%. With 1 MBq sodium-22 source, the signal rate is 0.01 Hz, including HFS transition probability, positronium formation rate of 20% in nitrogen and o-Ps ratio (75%). The accidental background is 0.014 Hz with accidental tagging rate of 4.6% (200 ns timing window) and the o-Ps background is 0.09 Hz. The S/N ratio is $0.01/(0.014 + 0.09) = 0.096$, and significance of 30 σ should be obtained in 10^6 sec live time. Background subtraction can be done by comparing with data without gyrotron power.

5. Summary

We are preparing the first experiment probing Ps-HFS with direct method to investigate discrepancy of Ps-HFS value between theory and experiment. We have developed a high power 203 GHz radiation source gyrotron and a Fabry-Perot cavity with a metal-mesh mirror and a Cu concave mirror. Monte-Carlo simulation of the detection system shows that the observation of Ps-HFS is feasible. We are now assembling the detection system, and the first data taking is scheduled to start in this autumn.

References

[1] Mills A P Jr and Bearman G H 1975 *Phys. Rev. Lett.* **34** 246, Mills A P Jr 1983 *Phys. Rev.* A **27** 262
[2] Ritter M W, Egan P O, Hughes V W and Woodle K A 1984 *Phys. Rev.* A **30** 1331
[3] Kniehl B A and Penin A A 2000 *Phys. Rev. Lett.* **85** 5094

[4] Idehara T, Ogawa I, Mitsudo S, Pereyaslavets M, Nishida N and Yoshida K 1999 *IEEE Trans. Plasma Sci.* **27** 340
[5] Idehara T, Saito T, Ogawa I, Mitsudo S, Tatematsu Y, La Agusu, Mori H and Kobayashi S 2008 *Appl. Magn. Reson.* **34** 265

Journal of Physics: Conference Series **199** (2010) 012003 doi:10.1088/1742-6596/199/1/012003

Ps^- emission from Cs coated surfaces

K. Michishio[1], **T. Tachibana, H. Terabe, A. Miyamoto and Y. Nagashima**[2]

Department of Physics, Tokyo University of Science, Shinjuku, Tokyo 162-8601, Japan

E-mail: [1]j1208665@ed.kagu.tus.ac.jp, [2]ynaga@rs.kagu.tus.ac.jp

Abstract. Recently, we have observed the efficient emission of positronium negative ions from Cs coated W(100) and polycrystalline Mo and Ta surfaces. The efficient emission enables us to perform new experimental studies on these ions. The investigations of the efficient production of positronium negative ions are reviewed and a measurement of the binding energy of the ions and a photodetachment experiment are proposed.

1. Introduction

The positronium negative ion (Ps^-), a bound state of a positron and two electrons, is the simplest three-body system which provides a good sample for testing QED corrections. Therefore, a number of theoretical investigations have been carried out since the first prediction of the stability of this ion by Wheeler in 1946 [1]. For example, the binding energy E_B has been calculated by several authors [2]–[6] and resonances in the electron-positronium (Ps) continuum have been studied [7][8]. The photodetachment cross sections have also been calculated [9][10]. However, after the first observation of Ps^- by Mills [11] in 1981, no experimental investigations have been performed on this system except for a few measurements of the decay rate [12]–[15]. In order to accomplish more extensive investigations into the fundamental processes, new methods of generating more efficient Ps^- production had to be developed.

When slow positrons impinge on a metal surface, they lose their kinetic energy via processes such as phonon excitation or electron-hole pair creation and are thermalized. Some of these diffuse back to the surface and if energetically allowed, they contribute to the positron re-emission or Ps emission. In 2006, we observed the spontaneous emission of Ps^- from a polycrystalline tungsten surface as another channel [16]. The Ps^- emission efficiency (the fraction of incident slow positrons yielding Ps^-) was only 0.007%. In 2008, we succeeded in significantly increasing this value by obtaining Ps^- from a W(100) surface coated with Cs [17].

In the present paper, the efficient emission of Ps^- from alkali metal-coated surfaces will be reviewed and future experimental investigations will be proposed.

2. Efficient emission of Ps^- from a Cs coated W(100) surface

The effect of Cs coating on the Ps^- emission from metal surfaces has been discussed by Mills [11] and Laricchia [18][19]. Mills suggested that it becomes energetically favorable to form Ps^- if the electron work function of a surface is lowered by coating it with Cs. Laricchia suggested that good yields of Ps^- might be obtained at positron incident energies of a few eV at grazing angles.

The formation method of Ps^- in our experiments is based on the spontaneous emission of these ions from a Cs coated surface. The Ps^- is emitted from a tungsten surface spontaneously because the Ps^- work function, expressed as

$$\phi_{Ps^-} = \phi_+ + 2\phi_- - E_B \tag{1}$$

is negative, where ϕ_+ and ϕ_- are the positron and electron work functions, respectively. Here, using the chemical potential for the electron and positron, μ_+, μ_-, the values of ϕ_+ and ϕ_- can be written as

$$\phi_+ = -D - \mu_+, \ \phi_- = D - \mu_- \tag{2}$$

respectively, where D is the effect of the adsorbate dipole. Therefore, the Ps^- work function is expressed as

$$\phi_{Ps^-} = -\mu_+ - 2\mu_- - E_B + D. \tag{3}$$

Equation (3) shows that the Ps^- work function depends on the surface state. It is well known that alkali metal adsorption on a tungsten surface reduces the electron work function due to the depolarization of the surface dipole. As a result, the Ps^- work function can be decreased and hence the Ps^- emission efficiency will be increased.

The experimental system used was a magnetically guided slow positron beam apparatus with a trochoidal $\boldsymbol{E} \times \boldsymbol{B}$ filter. The beam intensity at the target was 1×10^5 e^+/s and the beam diameter was about 5mm.

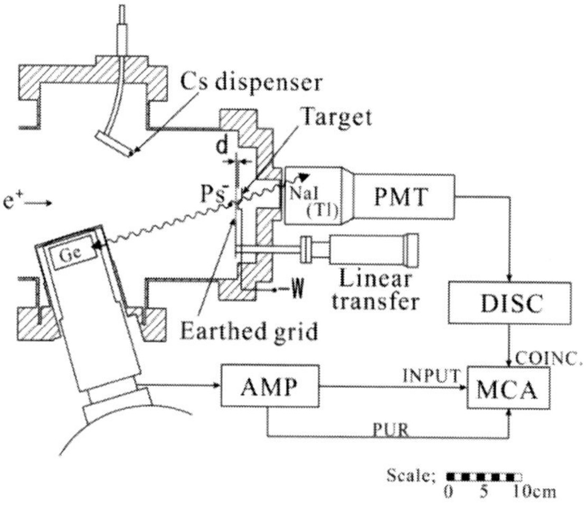

Figure 1. Schematic diagram of the target chamber [17].

Figure 1 shows the schematic diagram of the target chamber. The target was biased at -3kV so that the positrons were incident onto the target with an energy of 3.1keV. The Ps^- emitted from the target were accelerated by the electric field between the target and the grounded grid. A pure Ge detector was used to detect the blue-shifted annihilation γ-rays which were emitted on self-annihilation by the oncoming Ps^-. In order to detect the low count rate events of Ps^-, the background was reduced by using the Ge detector signal in coincidence with that from a NaI scintillation detector placed behind the target. The target was a W(100) film of thickness 2μm. It was annealed at 1800K *in situ* for 30 min and then coated with Cs using an alkali metal dispenser. The coverage of the Cs layer was set to be 2×10^{14} atoms cm^{-2}.

Figure 2 shows the annihilation photon energy spectra. The spectra for the unannealed and annealed targets were taken over a period of about 40h. The spectrum for the Cs coated surface was accumulated for 3000s immediately after deposition. The conversion efficiency for the uncoated W(100) after annealing was 0.005%. Figure 2 (c) shows the efficient emission of Ps^- from a Cs coated tungsten surface. After the coating, the emission efficiency, f, increased to 1.25%, which is 250 times higher than for the uncoated surface. The value decreased with time after deposition and was almost constant at 0.1% after 17h, which is still greater than that obtained using the beam-foil method.

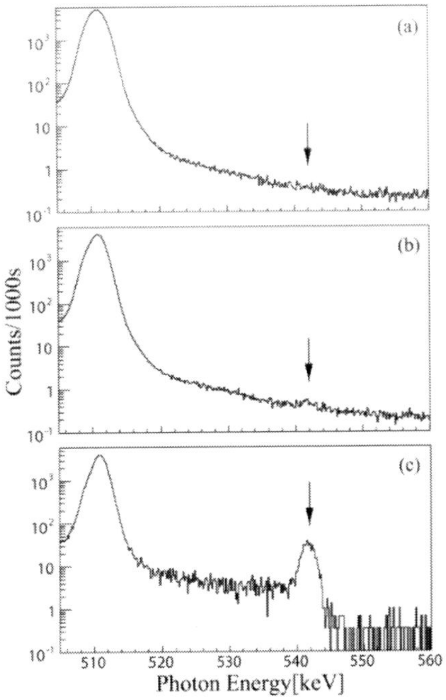

Figure 2. The annihilation photon energy spectra for (a) unannealed, (b) annealed and (c) Cs coated W(100) surfaces. The arrows indicate the photon energy of the self-annihilation of the accelerated Ps^-.

We have also tested coating with Na, which is less active chemically, in order to obtain long term stability. The coverage of the Na layer was estimated to be about 2×10^{14} atoms cm^{-2}. Surprisingly, immediately after the deposition, f was greater than that for the Cs coating and remained higher than 1% for more than one day.

3. Measurement of the Ps^- binding energy

In this work, we have made an estimate of the value of E_B, which has not been measured thus far, by observing the Ps^- emission from polycrystalline Mo and Ta surfaces. The targets were commercially obtained polycrystalline foils of 20μm thickness. They were annealed *in situ* at 1500°C for 30min. For the annealed Mo target, a low-intensity peak attributed to Ps^- emission was seen at 542 keV. For the Mo target with a 2×10^{14}atoms cm^{-2} Cs coverage, the value of f was 0.8%. This observation sets a lower limit to the Ps^- binding energy. The result indicates that the positron in the Mo target can leave the uncoated surface as Ps^-, i.e.,

$$\phi_+ + 2\phi_- - E_B < 0,$$

therefore,

$$E_B > \phi_+ + 2\phi_-.$$

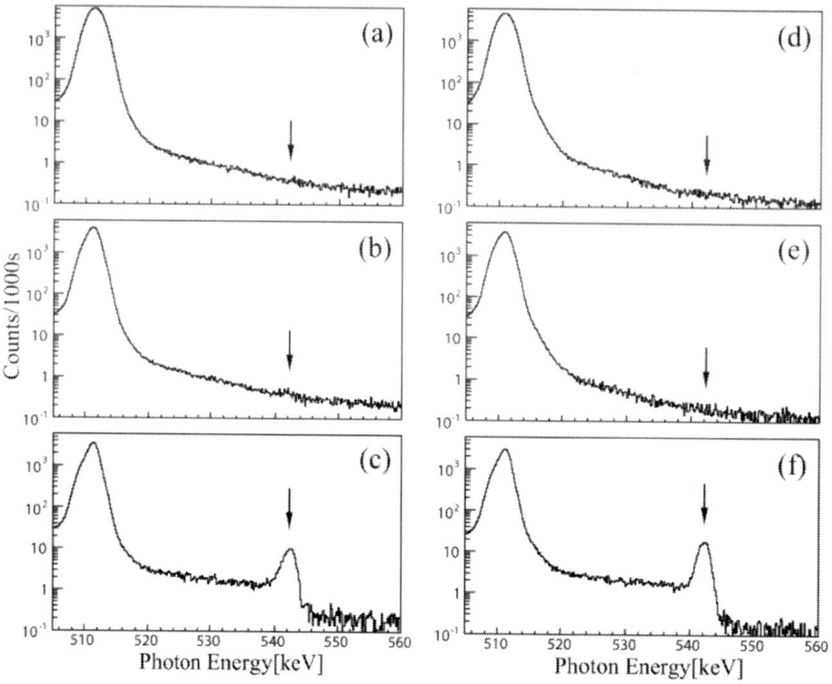

Figure 3. The annihilation photon energy spectra for (a) unannealed, (b) annealed and (c) Cs coated polycrystalline Mo surfaces and (d) unannealed, (e) annealed and (f) Cs coated polycrystalline Ta surfaces.

The right-hand term can be obtained from previous measurements of ϕ_+ [20] and ϕ_- [21] to be 7.1eV.

For the annealed Ta target, no significant peak was seen at 542keV though a peak due to Ps^- emission appeared in the spectrum after Cs coating. This indicates that the positrons in the target which diffuse back to the surface cannot leave the uncoated surface as Ps^-, i.e.,

$$\phi_+ + 2\phi_- - E_B > 0,$$

therefore,

$$E_B < \phi_+ + 2\phi_-,$$

which was determined to be obtained to be 7.3eV using previous measurements of ϕ_+ [22] and ϕ_- [21]. Therefore, E_B has been estimated to be

$$7.0\text{eV} < E_B < 7.3\text{eV}, \tag{4}$$

which is consistent with theoretical values that have been calculated [2][3][6]. This is the first experimental approach to estimate the binding energy of Ps^-. However, an accurate value is necessary to compare with the theoretical results.

In order to obtain such a value, we are planning to measure the dependence of the Ps^- yield on alkali metal coverage for the Ta target. We will search for the coverage corresponding to $\phi_{Ps^-} = 0$, and determine the binding energy as

$$E_B = \phi_+^0 + 2\phi_-^0, \tag{5}$$

where ϕ_+^0 and ϕ_-^0 are the positron and electron work functions measured for the coverage corresponding to $\phi_{Ps^-} = 0$, respectively [23].

Journal of Physics: Conference Series **199** (2010) 012003 doi:10.1088/1742-6596/199/1/012003

4. Ps^- photodetachment experiment

The significant increase obtained in the emission efficiency opens up the possibility of performing Ps^- photodetachment experiments. An experiment using the pulsed positron beam at KEK [24] together with a Nd: YAG laser synchronized to the positron beam is being planned [25]. Using the Ps^- photodetachment technique, the production of a variable energy Ps beam will be feasible [13][18].

5. Acknowledgment

We thank Professors T. Hyodo, A. Yagishita, T. Kuga, A. Igarashi, A. P. Mills, Jr. and G. Laricchia for valuable discussions and their encouragement. This work was supported by the Reimei Research Promotion project (Japan Atomic Energy Agency) and Matsuo Foundation.

References

[1] Wheeler J A 1946 *Ann. New York Acad. Sci.***48** 219
[2] Frost A A, Inokuti M and Lowe J P 1964 *J. Chem. Phys.* **41** 482
[3] Ho Y K 1993 *Phys. Rev.* A **48** 4780
[4] Frolov A M 1999 *Phys. Rev.* A**60** 1015
[5] Krivec R, Mandelzweig V B and Varga K 2000 *Phys. Rev.* A**61** 62503
[6] Drake G W F and Grigorescu M 2005 *J. Phys. B: At. Mol. Opt. Phys.* **38** 3377
[7] Ho Y K 1979 *Phys. Rev.* A**19** 2347
[8] Ho Y K 1984 *Phys. Lett.* A**102** 348
[9] Ward S J, Humberston J W and McDowell M R C 1987 *J. Phys. B: At. Mol. Opt. Phys.***20** 127
[10] Igarashi A, Shimamura I and Toshima N 2000 *New J. Phys.***2** 17
[11] Mills Jr. A P 1981 *Phys. Rev. Lett.***46** 717
[12] Mills Jr. A P 1983 *Phys. Rev. Lett.***50** 671
[13] Mills Jr. A P, Friedman P G and Zuckerman D M 1990 *Annihilation in Gases and Galaxies* ed R J Drachman (NASA, Maryland) p.213
[14] Schwalm D, et al. 2004 *Nucl. Instrum Methods* B **221** 185
[15] Fleischer F, et al. 2006 *Phys. Rev. Lett.***96** 063401
[16] Nagashima Y and Sakai T 2006 *New J. Phys.***8** 319
[17] Nagashima Y, Hakodate T, Miyamoto A and Michishio K 2008 *New J. Phys.***10** 123029
[18] Laricchia G 1995 *Positron Spectroscopy of Solids* ed A Dupasquier and A P Mills, Jr. (IOS, Amsterdam) p.401.
[19] Laricchia G, Armitage S, Kövér Á and Murtagh D J 2008 *Advances in Atomic, Molecular and Optical Physics* 56 ed E Arimondo, P R Berman and C C Lin (Elsevier, Amsterdam) p.1
[20] Jibary M, Weiss A, Koymen A R, Mehl D and Stiborek L 1991 *Phys. Rev.* B**44** 12166
[21] Lide D R (Ed) 2004 *CRC Handbook of Chemistry and Physics*, (CRC Press, Boca Raton)
[22] Weiss A H and Coleman P G 2000 *Positron Beams and Their Applications* ed P G Coleman (World Scientific, Singapore) p.129
[23] Nagashima Y, Hakodate T, Miyamoto A, Michishio K and Terabe H *Physicia Status Solidi* submitted
[24] Kurihara T, Yagishita A, Enomoto A, Kobayashi H, Shidara T, Shirakawa A, Nakamura K, Saito H, Inoue K, Nagashima Y, Hyodo T, Nagai Y, Inoue Y, Kogure Y and Doyama M 2000 *Nucl. Instr. and Method* B**171** 164
[25] Nagashima Y, Tachibana T, Michishio K, Terabe H, Hyodo T, Wada K, Yagishita A, Kurihara T and Kuga T 2009, private communication

Precise measurement of HFS of positronium

A Ishida[1], G Akimoto[1], K Kato[1], T Suehara[1], T Namba[1], S Asai[1], T Kobayashi[1], H Saito[2], M Yoshida[3], K Tanaka[3], A Yamamoto[3], I Ogawa[4], S Kobayashi[4] and T Idehara[4]

[1] Department of Physics and ICEPP, the University of Tokyo, 7-3-1 Hongo, Bunkyo-ku, Tokyo, 113-0033, Japan
[2] Institute of Physics, the University of Tokyo, 3-8-1 Komaba, Meguro-ku, Tokyo, 153-8902, Japan
[3] High Energy Accelerator Research Organization (KEK), 1-1 Oho, Tsukuba, Ibaraki, 305-0801, Japan
[4] FIR Center, University of Fukui, 3-9-1 Bunkyo, Fukui, 910-8507, Japan

E-mail: ishida@icepp.s.u-tokyo.ac.jp

Abstract. The ground state hyperfine splitting in positronium, $\Delta_{\rm HFS}$, is sensitive to high order corrections of QED. A new calculation up to $O(\alpha^3)$ has revealed a $3.9\,\sigma$ discrepancy between the QED prediction and the experimental results. This discrepancy might either be due to systematic problems in the previous experiments or to contributions beyond the Standard Model. We propose an experiment to measure $\Delta_{\rm HFS}$ employing new methods designed to remedy the systematic errors which may have affected the previous experiments. Our experiment will provide an independent check of the discrepancy. The measurement is in progress and a preliminary result of $\Delta_{\rm HFS} = 203.399\pm0.029$ GHz (143 ppm) has been obtained. A measurement with a precision of $O(1)$ ppm is expected within a few years.

1. Introduction

Positronium (Ps), a bound state of an electron and a positron, is a purely leptonic system which allows for very sensitive tests of QED. The precise measurement of the hyperfine splitting between orthopositronium (o-Ps, 1^3S_1) and parapositronium (p-Ps, 1^1S_0) (Ps-HFS) provides a good test of bound state QED. Ps-HFS is expected to be relatively large (for example compared to hydrogen HFS) due to a relatively large spin-spin interaction, and also due to the contribution from vacuum oscillation (o-Ps $\rightarrow \gamma^* \rightarrow$ o-Ps). The contribution from vacuum oscillation is sensitive to new physics beyond the Standard Model.

Figure 1 shows the measured and theoretical values of Ps-HFS. The combined value from the results of the previous 2 experiments is $\Delta_{\rm HFS,exp} = 203.388\,65(67)$ GHz (3.3 ppm) [1, 2]. Recent developments in nonrelativistic QED (NRQED) have added $O(\alpha^3)$ corrections to the theoretical prediction which now stands at $\Delta_{\rm HFS,th} = 203.391\,69(41)$ GHz (2.0 ppm) [3]. The discrepancy of 3.04(79) MHz (15 ppm, $3.9\,\sigma$) between $\Delta_{\rm HFS,exp}$ and $\Delta_{\rm HFS,th}$ might either be due to the common systematic uncertainties in the previous experiments or to new physics beyond the Standard Model.

There are two possible common systematic uncertainties in the previous experiments. One is the unthermalized o-Ps contribution which results in an underestimation of the material effect. This effect has already been shown to be significant [4]. The other is the uncertainty in the magnetic field uniformity which was cited as the most significant systematic error by previous experimenters.

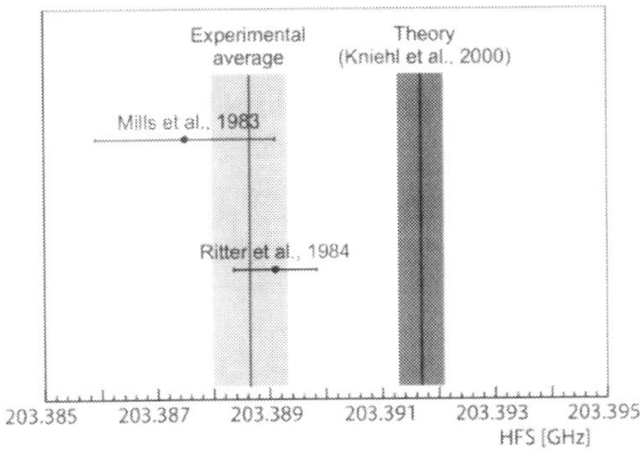

Figure 1. Measured and theoretical values of Ps-HFS.

2. Experimental setup

The energy levels of the ground state of Ps are shown as a function of static magnetic field in Figure 2. Due to technical difficulties in directly stimulating Δ_{HFS}, we make an indirect measurement by stimulating the transition Δ_{mix}. This is the same approach as previous experiments. The relationship between Δ_{HFS} and Δ_{mix} is given by the Breit-Rabi equation

$$\Delta_{\text{mix}} = \frac{1}{2}\Delta_{\text{HFS}}\left(\sqrt{1+x^2}-1\right), \quad (1)$$

in which $x = 2g'\mu_B H/h\Delta_{\text{HFS}}$. $g' = g\left(1-\frac{5}{24}\alpha^2\right)$ is the g factor for a positron (electron) in Ps [5], μ_B is the Bohr magneton, H is the static magnetic field, and h is the Plank constant.

In a static magnetic field, the p-Ps state mixes with the $m_z = 0$ substate of o-Ps hence the latter state annihilates into 2 γ-rays with a lifetime of about 8 ns (with our experimental conditions). The $m_z = \pm 1$ substates of o-Ps annihilate into 3 γ-rays with a lifetime of about 140 ns. When a microwave field with a frequency of Δ_{mix} is applied, transitions between the $m_z = 0$ and $m_z = \pm 1$ substates of o-Ps are induced so that the 2 γ-ray annihilation rate increases. This increase is our experimental signal.

Following are the main improvements in our experiment which we expect will significantly reduce systematic errors present in previous experiments.

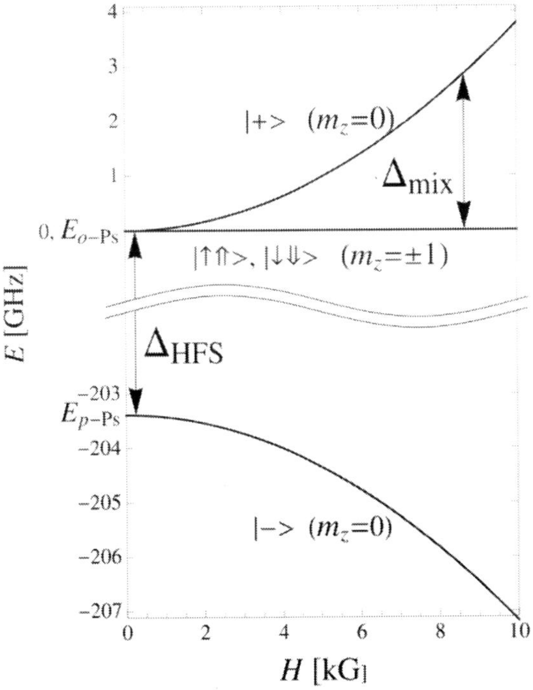

Figure 2. Zeeman energy levels of Ps in its ground state.

2.1. *Large bore superconducting magnet (Figure 3)*

A large bore superconducting magnet is used to produce the magnetic field which induces the Zeeman splitting (~ 0.866 T). The bore diameter of the magnet is 800 mm, and its length is 2 m. The large bore diameter means that there is good uniformity in the magnetic field in the region where Ps is formed (70 ppm at 0.61 G without utilization of any compensation).

Furthermore, the magnet is operated in persistent current mode, making the stability of the magnetic field better than 1 ppm.

2.2. *β-tagging system and timing information*

The positron source is 19 μCi (700 kBq) of ^{22}Na. A plastic scintillator 10 mm in diameter and 0.2 mm thick is used to tag positrons emitted from the ^{22}Na. The scintillation light

(a) The large bore superconducting magnet and the microwave waveguide. Microwaves are guided through the waveguide into the cavity.

(b) Inside of the magnet. The β-tagging system, the γ-ray detectors and the microwave cavity are located at the center of the magnet.

Figure 3. Photographs of the experimental setup.

is detected by photomultipliers and provides a start signal which corresponds to the time of Ps formation.

The positron then enters the microwave cavity, forming Ps in the N_2 gas contained therein.

Ps decays into photons that are detected with $LaBr_3$ (Ce) scintillators. Accumulating measurements of the times of positron emission and γ-detection results in decay curves of Ps as shown in Figure 4. The timing information is used to improve the accuracy of the measurement of Δ_{HFS} as follows:

(i) Imposing a time cut means that we can select well thermalized Ps, reducing the unthermalized o-Ps contribution. It should also be possible to precisely measure the contributions of unthermalized o-Ps, and of material effects (we plan to make such measurements in future runs).

(ii) A time cut also allows us to avoid the prompt peak (contributions of simple annihilation and of fast p-Ps decay), which greatly increases the S/N of the measurement.

2.3. *High performance γ-ray detectors*

Six γ-ray detectors are located around the microwave cavity to detect the 511 keV annihilation γ-rays. $LaBr_3$ (Ce) scintillators, 1.5 inches in diameter and 2 inches long are used. $LaBr_3$ (Ce) scintillators have good energy resolution (4 % FWHM at 511 keV) and timing resolution (0.2 ns FWHM at 511 keV), and have a short decay constant (25.6 ns). The good energy resolution and the high counting rate of $LaBr_3$ results in very good overall performance for measuring 2γ decays. In particular the good energy resolution allows us to efficiently separate 2γ events from 3γ events, negating the need to use a back-to-back geometry to select 2γ events, thus greatly increasing the acceptance of our setup.

This γ-ray detector system greatly reduces the statistical error in the measurement.

2.4. *RF system*

Microwaves are produced by a local oscillator signal generator and amplified to 500 W with a GaN amplifier (R&K A2856BW200-5057-R).

The microwave cavity is made with oxgen-free copper; the inside of the cavity is a cylinder 128 mm in diameter and 100 mm long. The side wall of the cavity is only 2 mm thick in order to allow the γ-rays to efficiently escape. The cavity is operated in the TM_{110} mode. The resonant frequency is 2.8532 GHz and $Q_L = 14700 \pm 50$. The cavity is filled with 1.5 atm gas (90 % N_2 and 10 %

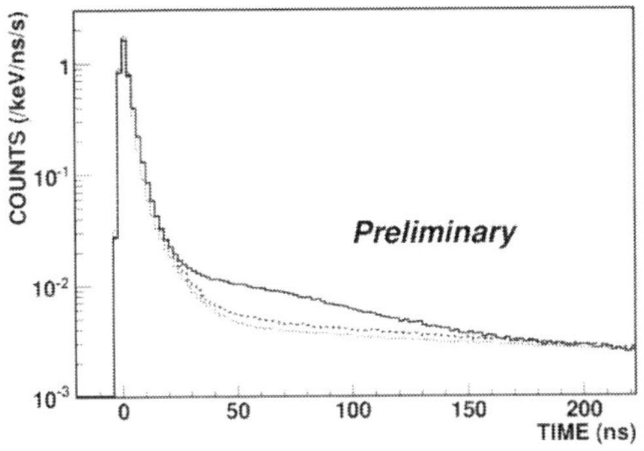

Figure 4. Decay curves of Ps at different magnetic field strengths. The solid line is 0.8659 T RF 450 W, the dashed line is 0.8614 T RF 450 W, and the dotted line is RF OFF. An energy window of 492–530 keV is applied.

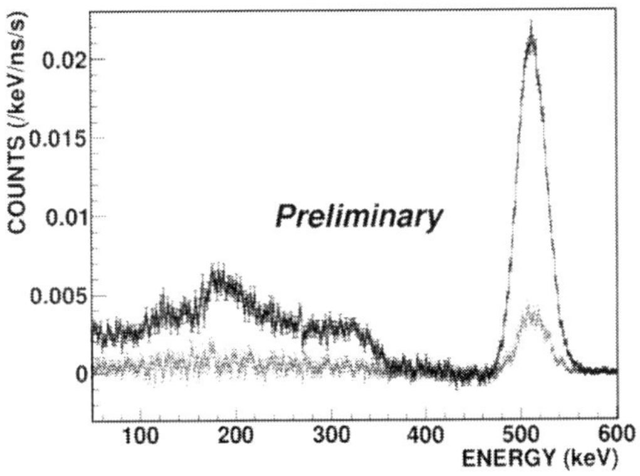

Figure 5. Energy spectra at different magnetic field strengths. The solid line is 0.8659 T RF 450 W and the dashed line is 0.8614 T RF 450 W. A timing window of 30–200 ns is applied, and the accidental spectra from 700–900 ns and RF OFF have been subtracted.

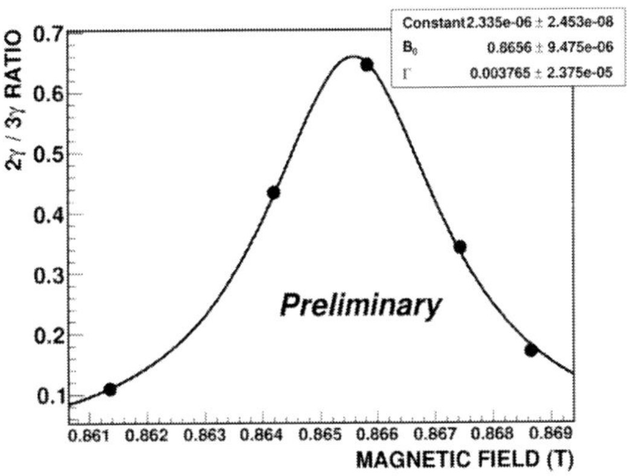

Figure 6. Resonance line at 1.5 atm. The data are fitted with the Breit-Wigner function (Equation (2)). Error bars are smaller than the marker size.

iso-C_4H_{10}) with a gas-handling system. Iso-C_4H_{10} is used as the quenching gas to remove background 2 γ-ray annihilations.

3. Analysis and current status

The first run of the experiment is ongoing using the large bore magnet with no compensation (compensation magnets to reduce the uniformity to $O(1)$ ppm are planned but are not yet installed). The 2 γ-ray annihilation rate has been measured at various magnetic field strengths with a fixed rf frequency.

3.1. Data analysis

Figure 4 shows examples of measured timing spectra. The peak coming from prompt annihilation and p-Ps decay is followed by the decay curve of o-Ps and then the constant accidental spectrum. A timing window of 30–200 ns is applied to select o-Ps events. Figure 5 shows the 2γ transition spectrum, which is obtained as follows:

(i) The accidental contribution is subtracted using the timing window t =700–900 ns.

(ii) The ordinary o-Ps decay spectrum is also subtracted. This spectrum is obtained by running without RF power, and is normalized to the 2γ transition spectrum in the region 380–460 keV.

The 2γ transition spectrum is measured at different magnetic field strengths. The resonance line obtained is shown in Figure 6.

Journal of Physics: Conference Series **199** (2010) 012004 doi:10.1088/1742-6596/199/1/012004

Fitting the measured points with the Breit-Wigner function,

$$f(B) = \frac{C}{(B - B_0)^2 + \Gamma^2/4}, \quad (2)$$

results in a constant of $C = 2.335(25) \times 10^{-6}$, a center value of the magnetic field of $B_0 = 0.865\,6151(99)$ T (11 ppm), and a width of $\Gamma = 3.765(24) \times 10^{-3}$ T.

The main source of systematic error is the uniformity of the magnetic field (142 ppm in Δ_{HFS}). Other sources are considered to be negligible. The preliminary value of Δ_{HFS} calculated from equation (1) is

$$\Delta_{\text{HFS}} = 203.399 \pm 0.005(\text{stat.}) \pm 0.029(\text{sys.})\ \text{GHz}\,(143\,\text{ppm}), \quad (3)$$

which is consistent with both of the previous experimental values and with the theoretical value.

3.2. Future plan

The following improvements are planned for future measurements:

(i) Compensation magnets will be installed and $O(1)$ ppm magnetic field uniformity is expected to be achieved.

(ii) Measurements at various pressures of N_2 will be performed to estimate the material effect (the Stark effect). The accumulation of these measurements will result in an $O(1)$ ppm statistical error within a few years.

(iii) The timing information allows for a measurement of Ps thermalization as a function of time [4]. We can thus precisely measure the material effect including the thermalization effect.

4. Conclusion

A new experiment to measure the Ps-HFS which reduces possible common uncertainties in previous experiments has been constructed and the first run is ongoing. A preliminary value of $\Delta_{\text{HFS}} = 203.399 \pm 0.005(\text{stat.}) \pm 0.029(\text{sys.})$ GHz (143 ppm) has been obtained, which is consistent with both of the previous experimental values and with the theoretical calculation. Development of compensation magnets is underway with a view to obtaining $O(1)$ ppm magnetic field homogeneity for future runs. A new result with an accuracy of $O(1)$ ppm will be obtained within a few years which will be an independent check of the discrepancy between the present experimental values and the QED prediction.

References

[1] Ritter M W, Egan P O, Hughes V W and Woodle K A 1984 *Phys. Rev.* B **30** 1331
[2] Mills A P Jr 1983 *Phys. Rev.* A **27** 262
[3] Kniehl B A and Penin A A 2000 *Phys. Rev. Lett.* **85** 5094
[4] Kataoka Y, Asai S and Kobayashi T 2009 *Phys. Lett.* B **671** 219
[5] Grotch H and Hegstorm R A 1971 *Phys. Rev.* A **4** 59

Magnetised Positronium

D P van der Werf, C J Baker, D C S Beddows, P R Watkeys, C A Isaac, S J Kerrigan, M Charlton and H H Telle

Department of Physics, School of Physical Sciences, Swansea University, Singleton Park, Swansea SA2 8PP, UK

E-mail: D.P.van.der.Werf@Swansea.ac.uk

Abstract. Magnetised positronium is formed by impacting low energy positrons onto a gas covered target immersed in a magnetic field (B ≥ 1T). The resulting weakly bound positronium atoms subsequently travel some distance in an arrangement of Penning-type traps whereupon they can be field ionised. The remnant positrons are accumulated and then detected by forced annihilation on the target. The production efficiency of the magnetised atoms has been measured for different species of gases, gas layer thickness and the strength of the magnetic field. The positronium loss as a function of the distance travelled has been measured and is shown to be caused by the magnetron drift of the positronium atom.

1. Introduction

Several years ago Estrada *et al.* [1] observed that, when bombarding a target consisting of a gas covered surface with low energy positrons in a 5.3 T magnetic field, a small fraction of the positrons formed magnetised positronium. The maximum yield per incoming positron was found to be 2×10^{-3}, as measured by field ionisation in a Penning trap located about 6 cm from the target. More recently, Jelenković *et al.* [2], who used a similar technique to load a trap to study sympathetic cooling of positrons using laser-cooled beryllium ions, found a maximum yield (6×10^{-7}) around three orders of magnitude lower than Estrada's results. Although not yet explained, this dramatic discrepancy could be caused by the different targets used in the two experiments. The target surface in Estrada's work consisted of a repeatedly annealed tungsten crystal held at cryogenic temperatures which was covered by a layer of gas resulting from direct condensation onto the surface of residual gas species from their vacuum chamber following their cryogenic cool down procedure. They showed that when the gas was removed the yield plummeted. On the other hand, Jelenković *et al.* used an annealed room temperature copper target in a baked UHV system.

Both experiments showed that magnetised, and therefore loosely bound, positronium can be transported over some distance. Thus, such positronium atoms could possibly be used to form antihydrogen by exploiting a charge exchange process with an antiproton [3, 4, 5, 6]. This, and the fact that the role of gas on the target is not yet properly understood, prompted the research reported here.

In this paper we present the results of magnetised positronium studies using a number of gases and target coverages as a function of the magnetic field, the maximum electric field used for ionisation and the distance travelled before ionisation.

2. Experimental Details

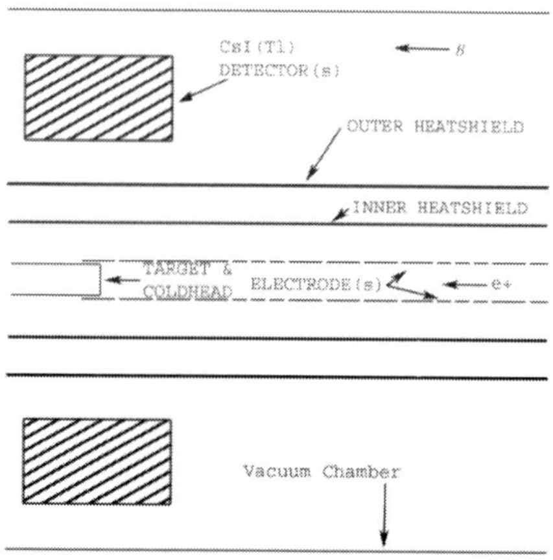

Figure 1. Schematic diagram of the target area.

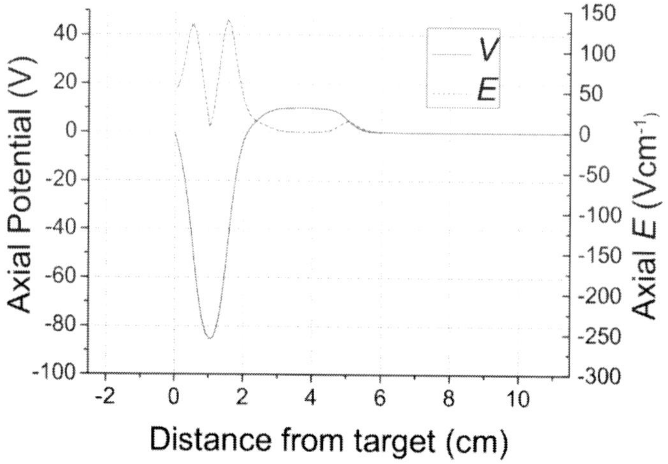

Figure 2. Typical electrical potential on axis (V) and the magnitude of the electric field E as calculated using Simion^{TM}.

Positrons originating from a 0.67 GBq ^{22}Na source were moderated using solid neon, resulting in a beam of $\sim 1 \times 10^6$ e^+s^{-1} [7]. These positrons were accelerated to an energy of about 100 eV, and were magnetically guided into an electrode array and implanted into a biased, gold plated, target (see figure 1). The target and electrodes were immersed in a high magnetic field (which could be varied in the range 1-5 T) and were able to reach cryogenic temperatures. To that effect the electrodes were in thermal contact with the inner heat shield (see figure 1) which was connected to, together with the electrically isolated (using sapphire) target, the second stage of a SHI-Cryogenics SRDK-408E cold head. The lowest temperatures reached were 6.7 K on top of the inner heatshield and 3.2 K on the second stage of the cold head as measured by calibrated CERNOX sensors.

Weakly bound positronium atoms, formed at the surface of the target, were ionised by an electric field similar to that shown in figure 2. The resultant free positron was then trapped in the potential well. After a certain "load time" the incoming beam was blocked and the positrons dumped back onto the target. The subsequent gamma rays were detected by the two in-situ CsI detectors shown in figure 1. Throughout this paper we will use the observable R, which is the ratio of the number of trapped positrons, obtained by field ionisation of the positronium atoms, per incoming positron per second.

3. Results and Discussion

Positronium can be said to be in a magnetised state by applying insights gained by Glinsky and O'Neil [8] (and used in our earlier work [9]) into the properties of the so-called guiding centre atoms. In that work a guiding centre atom is a weakly bound ion-electron pair immersed in a strong magnetic field. The pair exhibit three separate motions which can be described by three characteristic frequencies: the cyclotron motion with frequency $\omega_c = eB/m$, the axial motion, caused by the inter-particle potential, with a frequency $\omega_z = (e^2/(4\pi\epsilon_0 m r_{ee}^3))^{1/2}$ and the drift with a frequency $\omega_d = e/(8\pi\epsilon_0 B r_{ee}^3) = \omega_z^2/(2\omega_c)$. Here e and m are the electronic charge and electron/positron mass, respectively, r_{ee} is the distance between the positron and electron and B is the magnitude of the magnetic field. The pair is defined to be magnetised when

Journal of Physics: Conference Series **199** (2010) 012005 doi:10.1088/1742-6596/199/1/012005

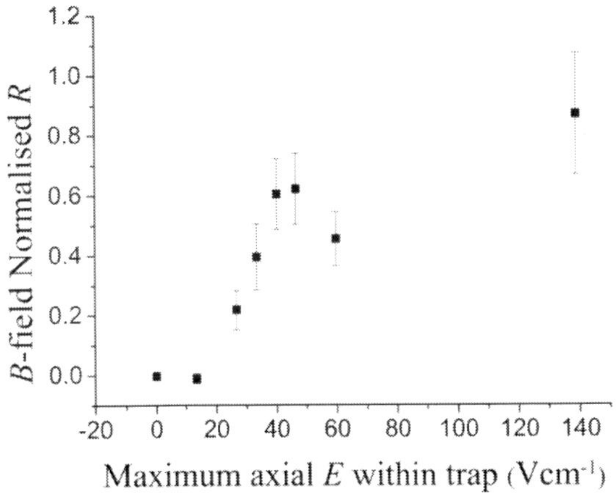

Figure 3. R as a function of the maximum ionisation field ("original measurements"). The ionisation well was 1 cm away from the target and the load time was 2000 s. The data are a weighted average for B = 1, 3 and 5 T, where the weighting was derived from figure 6.

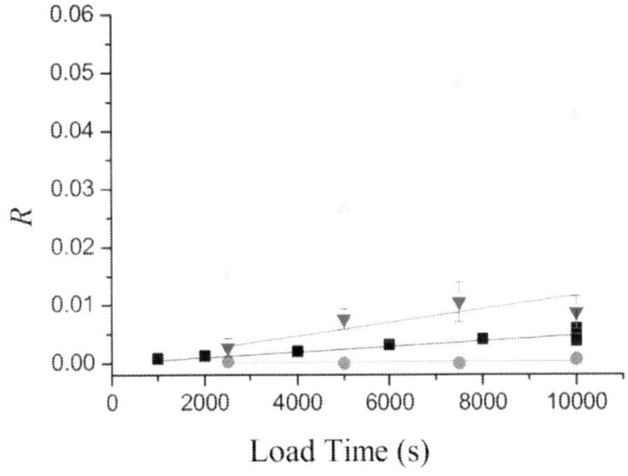

Figure 4. R as a function of the load time for ionisation wells at z = 1 cm (▲), 2 cm (▼), 3 cm (■), 8.5 cm (●). The data were taken under conditions in which the field used for ionisation was E = 139 V cm^{-1} and at B = 5 T. The lines represent linear fits to the data points.

$\omega_c >> \omega_z >> \omega_d$ which can be translated to the condition $\omega_c >> \omega_z/2$. Reworking leads to a condition for the inter-particle distance as,

$$r_{ee} >> \left(\frac{m}{16\pi c_0 B^2}\right)^{1/3}. \tag{1}$$

For magnetic fields between 5 and 1 T the above condition becomes $r_{ee} >> 0.04 - 0.13\ \mu$m.

An initial experiment, using a clean target surface (i.e. no gas deposited on it), established that the value of R was zero. Only after inserting nitrogen for extended periods via an upstream buffer gas accumulator device [7] whilst maintaining the target and electrodes at ~ 6 K, was the Rydberg positronium signal observed. No optimisation of this process took place. In the discussion that follows results obtained in this manner are called "original". Subsequently, the experimental apparatus was altered in such a way that gas could be admitted into the chamber in a controlled manner. Typically, a certain gas pressure was maintained for a specific time while the system was cooled down from just above the atmospheric condensation temperature of the gas to about 4.5 K. By varying the pressure of a particular gas the target was exposed to a larger amount of gas and, therefore, likely a higher coverage on the target was formed. It has been assumed that this coverage is proportional to the product of the pressure and the exposure time. However, it is unlikely that the coverage of the various gas species is the same for a given pressure × exposure time.

Figure 3 shows the effect of the maximum axial ionisation field on R. This parameter is essentially zero until about 20 Vcm^{-1}, then rises up to about 40 Vcm^{-1}, after which it stays approximately constant. The magnitude of the self electric field at the positron or electron is given by $e/(4\pi\epsilon_0 r_{ee}^2)$. When equating this to the maximum ionisation field we derive values of r_{ee} between 0.5 – 0.8 μm, indicating that the positronium atoms were just in the magnetised regime.

Figure 4 shows that R increases linearly with time indicating that the lifetime of the positrons in the trap is much in excess of 10000 seconds which is our maximum load time. The fitted

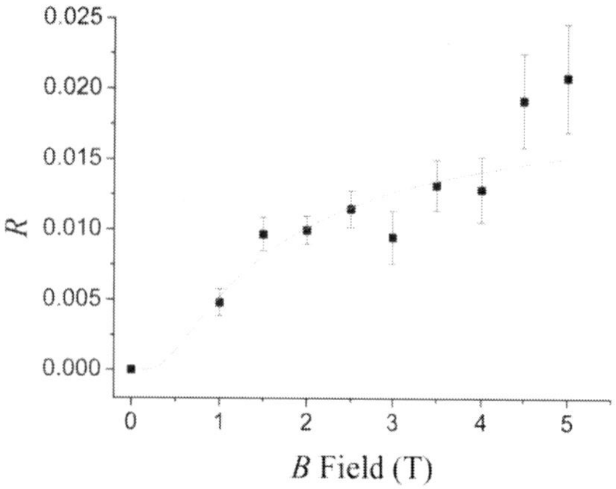

Figure 5. Gradient of the lines in figure 4 as a function of the distance from the target. The solid line is a fit of the functional form $R_t = Ae^{-z/z_c}$ (see text).

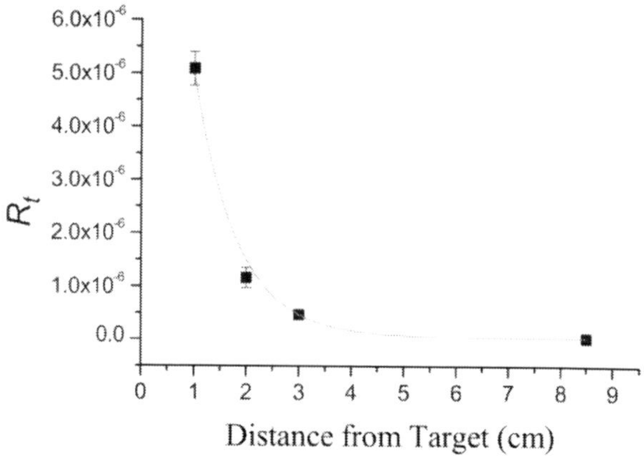

Figure 6. R as a function of B. The ionisation well is at $z = 1$ cm from the target and the maximum $E \sim 38$ Vcm^{-1}. The solid line is a fit of the form $R = Ae^{-D/B}$.

slopes in figure 4, which give the time normalised R, R_t, are plotted as a function of the distance travelled by the positronium from the target to its position of field ionisation (see figure 5). The solid line is a fit given by $R_t = Ae^{-z/z_c}$ with the amplitude, A, and the critical distance, z_c, being free parameters.

The exponential decrease of R_t can be explained by realising that magnetised positronium will have a transverse (magnetron) drift speed, v_d, brought about by the $\mathbf{E} \times \mathbf{B}$ drift, where the relevant electric field is that between the positron and the electron. A positronium atom at a position z will have travelled for a time $t = z/\langle v_{Ps}\rangle$, where $\langle v_{Ps}\rangle$ is the average axial positronium speed and will have moved a transverse distance $r = v_d t = v_d z/\langle v_{Ps}\rangle$. So, $z_c = r_c\langle v_{Ps}\rangle/v_d$, where r_c is the critical radius and is defined by the electrode geometry. This radius should be smaller than, or equal to, the inner trap electrode radius, $r_{el} = 5$mm. Using $v_d = \omega_d r_{ee}$ we derive that $z_c \lesssim \langle v_{Ps}\rangle r_{el} B r_{ee}^2/720$ m, where r_{ee} is given in μm. The fitted line in figure 5 gives a value for $z_c \approx 8$ mm. Therefore, $\langle v_{Ps}\rangle r_{ee}^2 \gtrsim 280$ ms$^{-1}(\mu\text{m})^2$, which leads to a positronium temperature lower than the 3.2 K target temperature when assuming the values of r_{ee} derived above. The data points in figure 5 can also be fitted with the function Az^n, resulting in a value of $n = -2.0 \pm 0.3$ indicating isotropic emission of the positronium atoms. However, this is inconsistent with the conclusion derived above that the positronium is magnetised.

In figure 6 R is plotted as a function of the magnetic field for a well at 1 cm distance and a maximum ionisation field of about 38 V cm^{-1}. Combining the expression above for z_c and the function for R used in figure 5 we can show that $R = Ae^{-D/B}$, where A and D are constants. The fitted line in figure 6 together with the above mentioned value for r_{el} gives a value for $\langle v_{Ps}\rangle r_{ee}^2 \sim 1.3 \times 10^3$ ms$^{-1}(\mu\text{m})^2$, which is not inconsistent with the conclusion derived from figure 5.

Using the data provided by the paper of Estrada *et al.*, where $r_{ee} \sim 2$ μm, $B = 5.3$ T and assuming a critical distance $z_c \sim 6$ mm (which is the distance between their target and ionisation well) and $r_{el} \sim 3.5$ mm a value for the mean positronium speed of about 600 ms^{-1} can be derived. This is comparable to the low positronium temperature found above with the caveat that the speed is dependent on the value of z_c, which was not explicitly given by Estrada *et al.*

Figure 7 shows the result of an investigation of the effect of target bias voltage on the value of R. Contrary to Estrada *et al.*, who obeserved a marked dependence of their Rydberg positronium yield on the applied voltage, our measured yield is essentially constant between -2 V and 1 V.

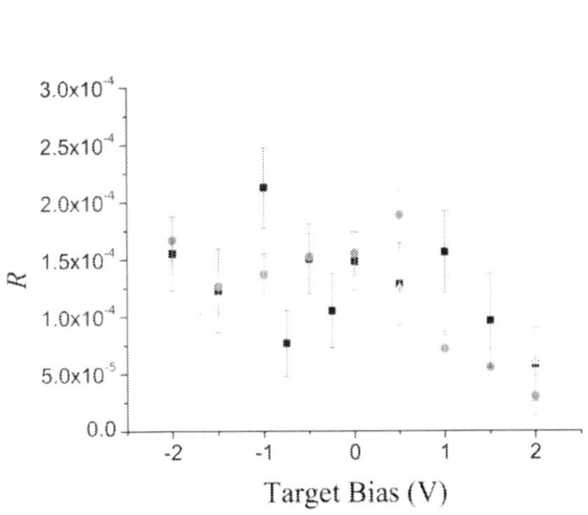

Figure 7. R as a function of the bias voltage on the target for O_2 (■), Ne (●), N_2 (▲). The ionisation well is at $z = 1$ cm.

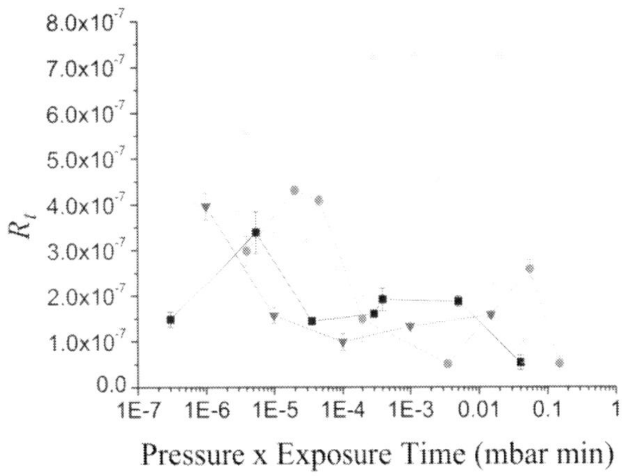

Figure 8. R_t as a function of the pressure times exposure time of the target for CO (■), N_2 (●), O_2 (▲), Ne (▼), dry air (◆). $|E| = 139$ V cm^{-1} and the ionisation well is at $z = 1$ cm.

It is also noticeable that, to within the uncertainties, the yield is species independent.

Figure 8 shows the effect on R as a function of the pressure × exposure time (see section 2) for a number of different gas species. It is clear that for each gas the yield depends on the amount of gas admitted. However, there are insufficient (properly calibrated) data to draw any firm conclusions.

4. Conclusions

Yields of R have been measured for a number of gas species as a function of B, distance from the target and the amount of admitted gas and have been found to be independent of gas species and much closer to the values found by Jelenković *et al.*, rather then those of Estrada *et al.* The origin of these discrepancies probably lies in the differences in target surface properties. In order to resolve the large difference in yield more work with different types of target and gases is needed.

The behaviour of the yield versus distance from the target has been adequately explained by the $\mathbf{E} \times \mathbf{B}$ drift caused by the fact that the positronium is magnetised. The outcome of the analysis of this drift motion implies that the Rydberg positronium atoms are emitted from the target with relatively low axial speed.

There is an indication that the amount of gas deposited on the target influences the yield, but more work, with better characterised surface coverages, is needed.

5. Acknowledgements

We would like to thank the EPSRC for their support (grant GR/81541/01) and the studentships awarded (CJB, CAI, SJK, PRW). We are grateful for the the technical support provided by the Swansea Physics Department.

[1] Estrada J, Roach T, Tan J N, Yesley P and Gabrielse G 2000 *Phys. Rev. Lett.* **84** 859
[2] Jelenković B M, Newbury A S, Bollinger J J, Itano W M and Mitchell T B 2003 *Phys. Rev. A* **67** 063406
[3] Deutch B I, Charlton M, Holzscheiter M H, Hvelplund P, Jørgensen L V, Knudsen H, Laricchia G, Merrison J P and Poulsen M R 1993 *Hyperfine Interact.* **76** 153
[4] Charlton M 1990 *Phys. Lett. A.* **143** 143
[5] Hessels E A, Homan D M and Cavagnero M J 1998 *Phys. Rev. A* **57** 1668

[6] Storry C H, Speck A, Sage D L, Guise N, Gabrielse G, Grzonka D, Oelert W, Schepers G, Sefzick T, Pittner H, Herrmann M, Walz J, Hänsch T W, Comeau D and Hessels E A 2004 *Phys. Rev. Lett.* **93** 263401
[7] Clarke J, van der Werf D P, Griffiths B, Beddows D C S, Charlton M, Telle H H and Watkeys P R 2006 *Rev. Sci. Inst.* **77** 063302
[8] Glinsky M E and O'Neil T M 1991 *Phys. Fluids B* **3** 1279
[9] Baker C J, van der Werf D P, Beddows D C S, Watkeys P R, Isaac C A, Kerrigan S J, Charlton M and Telle H H 2008 *J. Phys. B: At. Mol. Opt. Phys.* **41** 245003

Positronium formation cross-sections for Xe, CO_2 and N_2

D A Cooke, D J Murtagh and G Laricchia

UCL, Department of Physics and Astronomy, University College London, Gower Street, London, WC1E 6BT, UK

E-mail: g.laricchia@ucl.ac.uk

Abstract. The positronium formation cross-sections for Xe, CO_2 and N_2 have been measured using coincidences between γ-rays from positronium self-annihilation and the resultant ion. In the case of Xe, there is excellent agreement with previous experimental determinations. For CO_2 there is broad agreement in magnitude with previous measurements in contrast with N_2 where good shape agreement at low energies (< 40 eV) is found though the magnitude of the present cross-section is significantly higher.

1. Introduction

When considering ionization by positron impact, there are two dominant channels: direct ionization ($e^+ + A \rightarrow A^+ + e^+ + e^-$) and positronium (Ps) formation ($e^+ + A \rightarrow A^+ + Ps$). For an atomic target, the total ionization cross-section (Q_i^t) is defined as the sum of the Ps formation (Q_{Ps}) and direct ionization (Q_i^+) cross-sections, contributions from higher order processes and from direct annihilation of a target electron are considered generally negligible [1]. This is illustrated in figure 1, which shows the partitioning of Q_i^t for He into contributions from Q_i^+ and Q_{Ps}. In the case of a molecular target, the dissociative ionization cross-section ($Q_i^{\rm diss}$) must also be included in the sum for Q_i^t.

Recently, there have been experimental determinations of Q_{Ps} for the noble gases [e.g. 2, 3] and the first experimental measurements of Ps formation into the $2P$ state ($Q_{Ps}(2P)$) for He, Ar and Xe [4]. With the exception of He, there have been fewer theoretical determinations of Q_{Ps}, however, there are a number of cross-sections for excited-state Ps formation [e.g. 5, 6] and Ps formation from inner-shell electrons [e.g. 6–8].

Figure 2 shows the available experimental determinations of Q_{Ps} and $Q_{Ps}(2P)$ for Xe compared with theory. There is a large distribution of magnitudes and shapes when considering all determinations. There is, however, convergence between two recent measurements, those of [2] and [3]. The determination of [16], measured using γ-ray–ion coincidences, reproduces the structure observed by [2] in energy dependence only—an absolute scale was set by normalizing the coincidence yield to this previous determination. These measurements also suffered from a systematic effect at high energies ($E > 16$ eV) whereby the magnetic field failed to contain scattered projectiles, yielding an excess of γ-rays from annihilation on the cell walls.

In the present work, γ-ray–ion coincidences have been used to measure Q_{Ps} for Xe, N_2 and CO_2 with particular attention paid to confining all positrons after scattering. Unlike previous work [15], an absolute scale has been set on these measurements using Q_i^t. For Xe, there is good agreement between all determinations [see e.g. 17]; for the molecules, a concurrent measurement of Q_i^t for Ar was used for normalization.

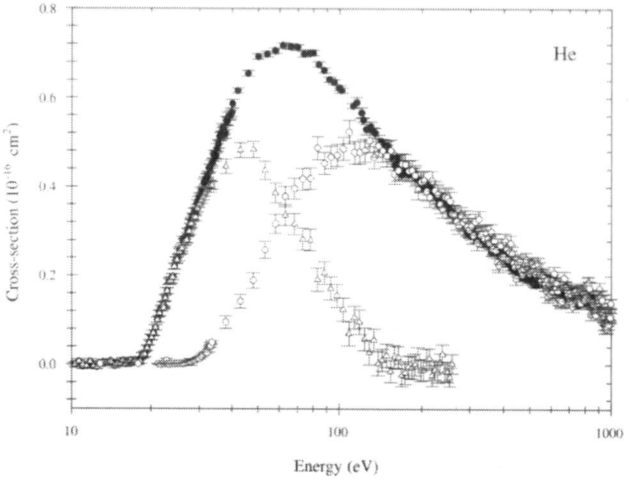

Figure 1. Partitioning of Q_i^t (●) of He into Q_i^+ (○) and Q_{Ps} (△). Data from [9–11].

2. Experimental Method

The experimental apparatus used in the present work is shown in figure 3 and has been described in detail elsewhere [18, 19]. Briefly, fast β^+ particles emitted from a ^{22}Na source are moderated by annealed W meshes producing a slow positron beam with $\Delta E \sim 2$ eV. The slow positrons are radially confined by a magnetic field ($\vec{B} \approx 100$ G) along the length of the beam-line (see black squares in figure 3). The beam passes through a bent solenoid, an electron repeller (repeller R1 in figure 3) and a Wien filter in order to reduce the number of unwanted γ-rays, secondary electrons and fast positrons transported to the interaction region. A positively-biased electrode (repeller R2 in figure 3) may be used to repel the slow portion of the beam, allowing measurement of the background produced by the remaining particles. The interaction region is a hemispherical gas cell constructed from polished Al. A small electrostatic lens held at -500 V extracts ions from the cell towards the detector, which consists of a channel electron multiplier (CEM) housed in a separately-pumped chamber. A CsI γ-ray detector is placed directly on top of the interaction region to detect annihilation quanta from the cell, and a second CEM is positioned at the end of the beamline for the detection of positrons. The photomultiplier tube mounted on an extension arm was used for the determination of $Q_{Ps}(2P)$.

The measurements were normalized by recording the total ion yield (Y_i)[1] simultaneously with the γ-ray–ion coincidence yield. Y_i is proportional to Q_i^t via:

$$Q_i^t = \frac{1}{nl}\frac{\varepsilon_+}{\varepsilon_i}Y_i \tag{1}$$

where n is the number density of the target gas, l is the effective cell length and ε refers to a detector efficiency. This normalization is performed similarly to that for Q_i^+ in [18], and allows the determination of $\frac{1}{nl}\frac{\varepsilon_+}{\varepsilon_i}$ for use in other normalizations. The error on this method is approximately $\pm 5\%$ [20]. The γ-ray–ion coincidence yield can then be normalized relative to the ion yield by correcting for the CsI detector efficiency which has been measured as 0.010 ± 0.001. This places an additional $\pm 10\%$ error on the absolute scale of the measured cross-sections.

[1] Yield is defined as event per positron recorded

Journal of Physics: Conference Series **199** (2010) 012006 doi:10.1088/1742-6596/199/1/012006

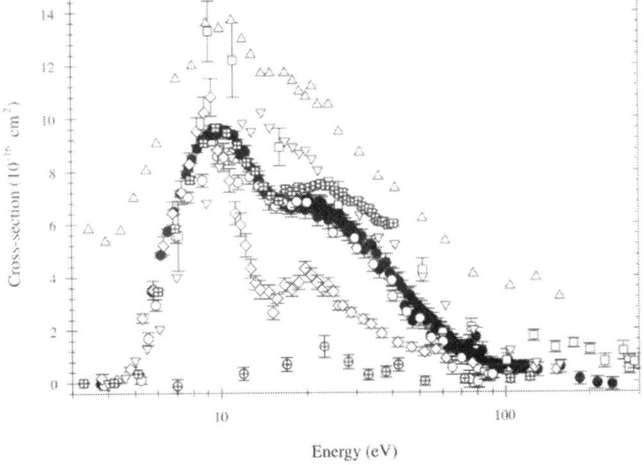

(a)

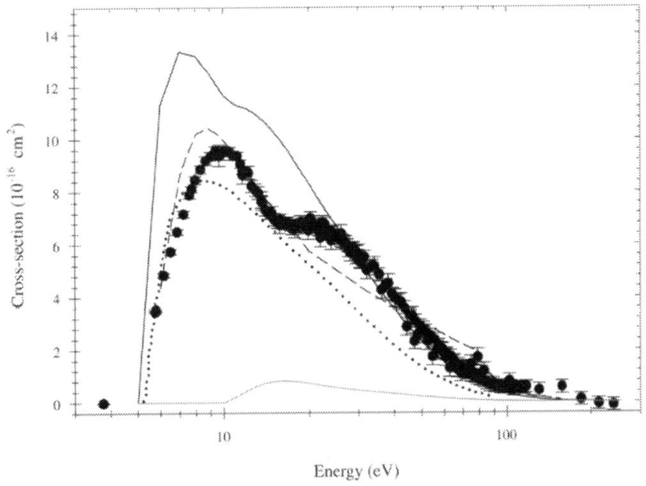

(b)

Figure 2. Review of experimental and theoretical determinations of Q_{Ps} for Xe. 2(a), experimental: ●—[2], ○—[3], ◇—[12], □—[13], ▽and △—lower and upper limits of [14] respectively, ⊞—[15], ⊕—$Q_{Ps}(2P)$ [4]. 2(b), theoretical: solid line—[6]×0.5, dashed line—[7]×0.5, dotted line—[8], solid grey line—$Q_{Ps}(2P)$ [6], ●—as 2(a).

3. Results

Figures 4, 5 and 6 show Q_{Ps} for Xe, CO_2 and N_2, respectively. In the case of Xe (figure 4), the present determination of Q_{Ps} peaks at ~10 eV with a magnitude of $8.9 \pm 0.3 \times 10^{-16}$ cm^2. A shoulder feature is observed between ~14 eV and 24 eV with the cross-section becoming negligible after ~100 eV. In comparison with other determinations of Q_{Ps}, there is excellent agreement, within errors, between the present results and those of [2] and [3].

Figure 5 shows the present Q_{Ps} for CO_2 compared with other experimental measurements. The present results peak at ~23 eV with a magnitude of $4.1 \pm 0.1 \times 10^{-16}$ cm^2 and extend to 750 eV. In

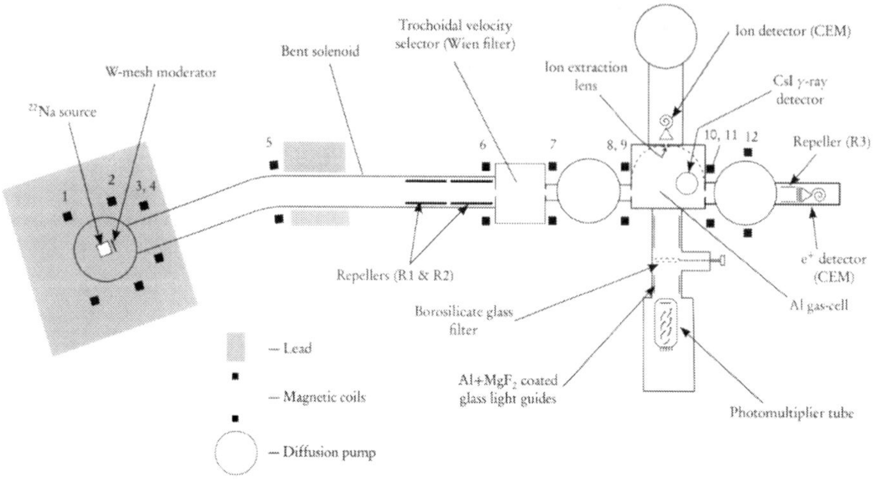

Figure 3. The magnetically guided positron beam and interaction region, showing the placement of various detectors.

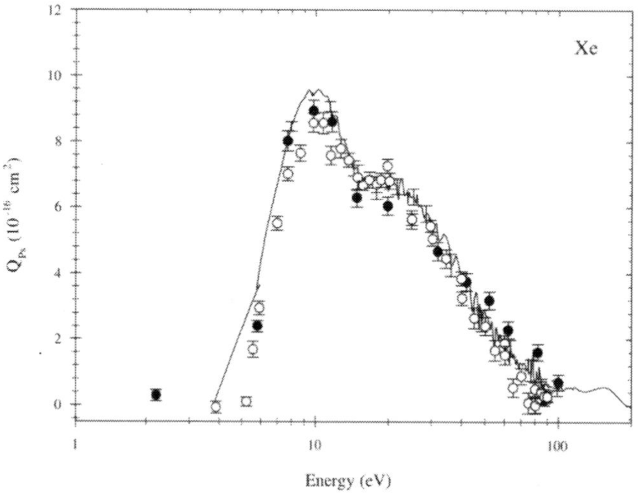

Figure 4. Determinations of Q_{Ps} for Xe: ●—present measurement, ○—[3], continuous curve—[2] (representative error bars are displayed).

comparison with previous determinations, the present Q_{Ps} displays fair agreement in the position and magnitude of the peak with the lower limit measurement of [22]. The present results are higher than the previous determination of [21] except in the peak region.

Finally, in the case of N_2 (see figure 6) the present measurement peaks at ~25 eV with a magnitude of $3.57 \pm 0.06 \times 10^{-16}$ cm^2 and extends to 350 eV. In comparison with the data of [23] there is excellent agreement in shape up to 40 eV (see inset), though the magnitude disagrees by $30 \pm 15\%$. Unlike for CO_2 [18], the separation of dissociation products from the parent ion has not been achieved for N_2. The extent to which N^+ is detected will affect the absolute scale placed on the reported results, and is the subject of ongoing investigation.

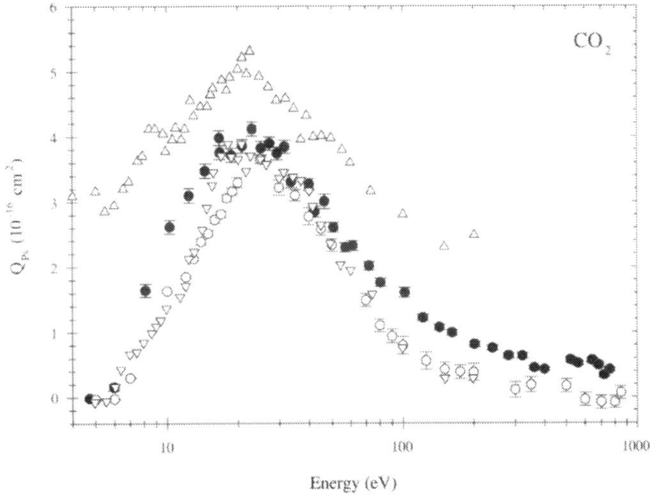

Figure 5. Q_{Ps} from CO_2: ●—present measurement, ○—[21], △ and ▽—upper and lower limits of [22], respectively.

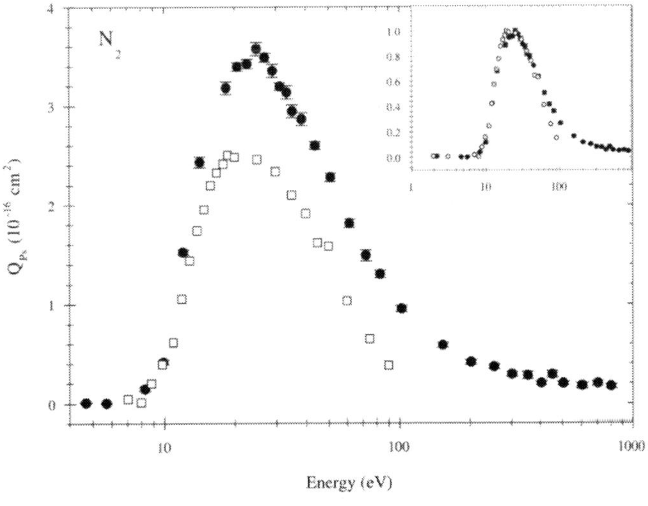

Figure 6. Q_{Ps} from N_2: ●—present measurement, □—[23]. Inset shows data normalized to each other for comparison of energy dependence (arbitrary units).

4. Conclusions

In the present work, γ-ray–ion coincidences have been used to determine Q_{Ps} for Xe, CO_2 and N_2. In the case of Xe, excellent agreement has been found between the present results and the two most recent determinations [2, 3]. For CO_2, there are a number of determinations of Q_{Ps} among which agreement is moderate. For N_2, there is an excellent agreement in shape between the present results and [23] up to 40 eV, although there is a $\sim$30% difference in magnitude. The apparent importance of Ps formation in molecules several hundred eV above threshold (in contrast to atoms) as implied by the present data is currently not understood, though it is noted that N_2 was found to be an efficient converter for positron-to-positronium beams up to $\sim$250 eV [24]. Clearly, further investigations are necessary.

5. Acknowledgements

We would like to thank J. Dumper and R. Jawad for technical assistance as well as EPSRC for funding this research.

References

[1] Van Reeth P, Laricchia G and Humberston J W 2005 *Physica Scripta* **71** C9
[2] Laricchia G, Van Reeth P, Szłuińska M and Moxom J 2002 *J. Phys. B* **35** 2525
[3] Marler J P, Sullivan J P and Surko C M 2005 *Phys. Rev. A* **71** 022701
[4] Murtagh D J, Cooke D A and Laricchia G 2009 *Phys. Rev. Lett.* **102** 133202
[5] Campbell C P, McAlinden M T, Kernoghan A A and Walters H R J 1998 *Nucl. Instr. Meth B* **143** 41
[6] Gilmore S, Blackwood J E and Walters H R J 2004 *Nucl. Instr. Meth B* **221** 129
[7] McAlinden M and Walters H 1992 *Hyperfine Interactions* **73** 65
[8] Dunlop L and Gribakin G 2006 *Nucl. Instr. Meth B* **247** 61
[9] Murtagh D J, Szłuińska M, Moxom J, Van Reeth P and Laricchia G 2005 *J. Phys. B* **38** 3857
[10] Ashley P, Moxom J and Laricchia G 1996 *Phys. Rev. Lett.* **77** 1250
[11] Moxom J, Ashley P and Laricchia G 1996 *Can. J. Phys* **74** 367
[12] Charlton M, Clark G, Griffith T C and Heyland G R 1983 *J. Phys. B* **16** L465
[13] Diana L M, Brooks D L, Coleman P G, Chaplin R L and Howell J P 1989 *Positron Annihilation* ed Dorikens-Vanpraet L, Dorikens M and Segers D (World Scientific) p 311
[14] Stein T S, Harte M, Jiang J, Kauppila W E, Kwan C K, Li H and Zhou S 1998 *Nucl. Instr. Meth B* **143** 68
[15] Szłuińska M and Laricchia G 2004 *Nucl. Instr. Meth. B* **221** 100
[16] Szłuińska M and Laricchia G 2004 *Nucl. Instr. Meth. B* **221** 107
[17] Laricchia G, Armitage S, Kövér Á and Murtagh D J 2008 *Adv. At. Mol. Opt. Phys.* vol 56 (Elsevier) pp 1–47
[18] Cooke D A, Murtagh D J, Kövér Á and Laricchia G 2008 *Nucl. Instr. Meth. B* **266** 466
[19] Szłuińska M, Van Reeth P and Laricchia G 2002 *J. Phys. B* **35** 4059
[20] Cooke D A, Murtagh D J and Laricchia G In preparation
[21] Murtagh D J, Arcidiacono C, Pešić Z D and Laricchia G 2006 *Nucl. Instr. Meth. B* **247** 92
[22] Kwan C K, Kauppila W E, Nazaran S, Przybyla D, Scahill N and Stein T S 1998 *Nucl. Instr. Meth. B* **143** 61
[23] Marler J P and Surko C M 2005 *Phys. Rev. A* **72** 062713
[24] Leslie D E, Armitage S and Laricchia G 2002 *J. Phys. B* **35** 4819

Detailed calculations on low-energy positron-hydrogen-molecule and helium-antihydrogen scattering

E A G Armour[†], J N Cooper[†], M R Gregory[†], S Jonsell[*], M Plummer[**] and A C Todd[†]

[†]School of Mathematical Sciences, University of Nottingham, Nottingham NG7 2RD, UK

[*]Department of Physics, University of Swansea, Swansea SA2 8PP, UK

[**]Computational Science and Engineering, STFC Daresbury Laboratory, Warrington WA4 4AD,UK

E-mail: edward.armour@nottingham.ac.uk

Abstract. In this paper, we consider two scattering processes: low-energy positron-hydrogen-molecule and helium-antihydrogen scattering. In the positron-hydrogen-molecule scattering calculations, we use the Kohn variational method to calculate $Z_{\rm eff}$, the number of target electrons available to the positron for annihilation. In the helium-antihydrogen scattering calculations, we use the Rayleigh-Ritz variational method to calculate a wave function for the leptons as a function of the distance between the helium and the antihydrogen. This is used, together with the associated nuclear wave function and the wave function for $\alpha\bar{p}$ + Ps^-, to calculate the cross section for the rearrangement reaction $\mathrm{He} + \bar{\mathrm{H}} \rightarrow \alpha\bar{p} + \mathrm{Ps}^-$, using the T-matrix and a form of the distorted wave approximation. For both processes, positron-electron correlation is taken into account accurately using Hylleraas-type functions.

1. Introduction

The two scattering processes to be considered are positron-hydrogen-molecule and helium-antihydrogen scattering, both at low energies. These systems both contain two nuclei and the same leptons, namely a positron and two electrons. The calculations are carried out with the nuclei in a fixed position. Thus in the two scattering processes, we treat the motion of the same three light particles. As a consequence, the calculations on the two processes have many features in common.

We will begin by considering positron-hydrogen-molecule scattering. Attention will be focussed on the calculation of $Z_{\text{eff}}(k)$, the effective number of electrons in the H_2 target that are available to the incident positron for annihilation. k is the wave number of the positron. The positron annihilation rate, λ, is given by

$$\lambda = \pi r_0^2 c D Z_{\text{eff}}(k), \tag{1}$$

where r_0 is the classical radius of the electron and D is the density of H_2 molecules in the vicinity of the positron.

Another area of great current interest is the preparation of cold antihydrogen ($\bar{\text{H}}$) and the study of its properties by the ALPHA and ATRAP collaborations at CERN [1, 2]. It is of interest to determine the interaction of antihydrogen with simple atoms and molecules. In view of this, we have carried out calculations of very low-energy H$\bar{\text{H}}$ scattering [3] and, more recently, of He$\bar{\text{H}}$ scattering [4]. In this paper, we describe calculations that we have carried out of cross sections for the rearrangement reaction

$$\text{He} + \bar{\text{H}} \rightarrow \alpha\bar{p} + \text{Ps}^-, \tag{2}$$

at very low energies.

Unless otherwise states, the units used in these calculations are hartree atomic units.

2. Positron-hydrogen-molecule scattering

2.1. Calculations using the complex Kohn method

The coordinate system used in our calculations on e^+H_2 scattering is illustrated in Fig 1. The protons are fixed at A and B, which have coordinates $(0, 0, -\frac{R}{2})$ and $(0, 0, \frac{R}{2})$, respectively, where R is the internuclear distance. The prolate spheroidal coordinates (λ, μ, ϕ) used in the calculations are of the form

$$\lambda = \frac{r_A + r_B}{R} \quad \text{and} \quad \mu = \frac{r_A - r_B}{R}.$$

ϕ is the usual azimuthal angle of spherical polar coordinates. Unless stated otherwise, the internuclear distance R was fixed at $1.4a_0$, the equilibrium value for the H_2 molecule.

The Kohn trial function was taken to be of the form

$$\Psi_{\text{trial}} = \psi_p \phi_{\text{target}} + \sum_{i=0}^{n} g_i \chi_i. \tag{3}$$

The open channel function ψ_p was taken to have the form

$$S + a_t T$$

where

$$T = S + iC \tag{4}$$

$$S = N\frac{\sin[c(\lambda_1 - 1)]}{\lambda_1 - 1} \tag{5}$$

$$C = N\frac{\cos[c(\lambda_1 - 1)]}{\lambda_1 - 1} \quad (1 - \exp[-\gamma(\lambda_1 - 1)]) \tag{6}$$

where λ_1 is the ellipsoidal prolate spheroidal coordinate for the positron,

$$c = \tfrac{1}{2}kR,$$

N is a normalisation constant and ϕ_{target} is the H_2 target wave function. This is the simplest form for the function, ψ_p. It represents the s-wave component of the part of the overall scattering wave function that is of $\sum_g^+$ symmetry. $\{\chi_i\}_{i=0}^n$ are short-range correlation functions and a_t and $\{g_i\}$ are linear parameters to be determined by the complex Kohn method.

The short-range correlation functions are of the form

$$\chi_0 = C\exp[-\gamma(\lambda_1 - 1)]\phi_{\text{target}} \tag{7}$$

$$\chi_i = S_{23}\bar{N}\lambda_1^{a_i}\lambda_2^{b_i}\lambda_3^{c_i}\mu_1^{d_i}\mu_2^{e_i}\mu_3^{f_i}C(s_i,t_i)\exp[-\alpha\lambda_1 - \beta(\lambda_2+\lambda_3)]\Omega, \quad (i \neq 0), \tag{8}$$

where particle 1 is the positron and particles 2 and 3 the electrons, S_{23} is the electron symmetriser, α and β are positive, non-linear parameters and $\bar{N}$ is a normalisation constant. $a_i, b_i, c_i, d_i, e_i, f_i$ are non-negative integers. The forms used for $C(s_i, t_i)$ are given in Table 1. If the method of models [5] is used

$$\Omega = \phi_{\text{target}}.$$

Otherwise Ω is taken to be equal to one.

Earlier calculations by the generalised Kohn method used the method of models [6]. In this method, the target Hamiltonian, $\hat{H}_t$, is taken to be

$$\hat{H}_t = -\tfrac{1}{2}\nabla_2^2 - \tfrac{1}{2}\nabla_3^2 + V_{\text{mod}} \tag{9}$$

where V_{mod} is such that

$$\hat{H}_t\phi_{\text{target}} = E_t\phi_{\text{target}} \tag{10}$$

where E_t is the expectation value of the energy of ϕ_{target}. No knowledge of V_{mod} is required when carrying out Kohn calculations. Also we do not have to consider the $\frac{1}{r_{23}}$ electron repulsion term in the exact form of the target potential.

This had the very considerable advantage that it was possible to include the key correlation functions in Table 1 containing $C(1,1) = \rho_{12} = \frac{2}{R}r_{12}$ without calculating integrals involving $\frac{\rho_{12}\rho_{13}}{\rho_{23}}$. These functions are Hylleraas-type functions that correlate the positron and the electrons. They play a very important role in improving the accuracy of the wave function at points at which the positron and an electron coincide that are the key to calculating $Z_{\text{eff}}(k)$ accurately. However, the complexity of the calculation only made it possible to use a target wave function that took into account 50% of the correlation energy of the H_2 molecule. This percentage is given by

$$\left(\frac{E_{\text{SCF}} - E_t}{E_{\text{SCF}} - E_{\text{ex}}}\right) \times 100 \tag{11}$$

where E_{SCF} is the SCF energy of H_2 and E_{ex} is the exact, non-relativistic energy of H_2 at $R = 1.4a_0$.

$Z_{\text{eff}}(k)$ is known to have a value of 14.6 for thermal positrons at 297K [7]. At this temperature, $kT = 26$ meV. Inclusion in the basis set of Hylleraas-type functions that correlate the positions of the positron and the electrons raised this to ~ 10.2 [6].

Our calculations [8] have been carried out without using the method of models. Inclusion of Hylleraas-type functions required the evaluation of integrals containing $\frac{\rho_{12}\rho_{13}}{\rho_{23}}$, in addition to those containing $\frac{\rho_{12}}{\rho_{13}}$ and $\rho_{12}\rho_{13}$ that were required when using the method of models. These evaluations were carried out using the Neumann expansion

$$\frac{1}{\rho_{ij}} = \sum_{\tau=0}^{\infty}\sum_{\nu=0}^{\tau} D_\nu^\tau P_\tau^\nu(\lambda_<)Q_\tau^\nu(\lambda_>)P_\tau^\nu(\mu_i)P_\tau^\nu(\mu_j)\cos[\nu(\phi_i-\phi_j)] \quad (12)$$

where $\lambda_<$ is the lesser of λ_i and λ_j and $\lambda_>$ is the greater of λ_i and λ_j, D_ν^τ is a coefficient and $P_\tau^\nu(x)$ and $Q_\tau^\nu(x)$ are the first and second solutions, respectively, to the associated Legendre equation.

The program used to evaluate the integrals containing $\frac{\rho_{12}\rho_{13}}{\rho_{23}}$ is to be published in *Computer Physics Communications* [9].

When carrying out the calculations without using the method of models, we were able to make the following improvements.

(1) We used a target wave function that took into account up to 99.7% of the correlation energy of H_2.

(2) The basis set used contained up to 297 short-range correlation functions as opposed to less than 100 in the calculation using the method of models.

(3) We included Hylleraas-type functions in the basis set to take into account electron-electron correlation. These were not included in the calculations using the method of models.

We used two target functions, ϕ_{tA} and ϕ_{tB}. ϕ_{tA} was calculated using 144 basis functions, none of which were Hylleraas-type functions. It took into account 96.8% of the correlation energy of H_2. ϕ_{tB} was calculated using the same 144 basis functions and one Hylleraas-type function containing ρ_{23} as a linear factor. It took into account 99.7% of the correlation energy.

The value of $Z_{\rm eff}(k)$ is given by the relation

$$Z_{\rm eff}(k) = \overline{\langle\Psi_{\rm trial}(k)|\sum_{i=2}^{3}\delta(\boldsymbol{r}_1-\boldsymbol{r}_i)|\Psi_{\rm trial}(k)\rangle} \quad (13)$$

where $\Psi_{\rm trial}(k)$ is the trial function obtained by the complex Kohn method for incident positrons with wave number k. $\Psi_{\rm trial}(k)$ is normalised so that the incident positrons have density one per unit volume. $\boldsymbol{r}_i$ is the position vector of particle i. The bar indicates averaging over all molecular orientations. The results obtained for $Z_{\rm eff}(k)$ at $k = 0.04$ are given in [8].

Recently, we have taken into account the motion of the nuclei in our calculations. As a first step we have determined how $Z_{\rm eff}(k)$ at $k = 0.04$ varies with the internuclear distance, R. The results obtained both with and without the use of the method of models are given in Table 2. It can be seen that the results obtained using the method of modules are larger than those obtained without using this method. The vibrationally averaged value of $Z_{\rm eff}(k)$, obtained using a Morse function, is 13.5 when obtained using the method of models and 9.7 without using this method [10].

The value obtained with the method of models is closer to the experimental value of 14.6 than that obtained without using it. In a very recent, very accurate calculation of $Z_{\rm eff}(k)$ at very low energies, Zhang *et al* [11] find similar behaviour to the values we obtain using the method of models as R increases beyond $1.6a_0$. It may be relevant that results obtained by the method of models have been found to have better convergence behaviour than those obtained without using this method [12, 8].

Journal of Physics: Conference Series **199** (2010) 012007 doi:10.1088/1742-6596/199/1/012007

2.2. Investigation of the cause of large $Z_{\text{eff}}(k)$ values

It is of interest to consider a model system which makes it possible to use the Kohn method to investigate how $Z_{\text{eff}}(k)$ might be greatly increased through resonances due to the existence of vibrationally excited quasi-bound states of the positron and the molecule. It is such states that are considered by Gribakin [13] to be responsible for the very large $Z_{\text{eff}}(k)$ values that have been observed in low-energy, positron scattering by some organic molecules [14].

No positron H_2 bound state exists. However, we can increase the 'positron' mass m_p to produce a weakly bound state.

Let us begin by applying the Born-Oppenheimer (B–O) approximation. The leptonic Hamiltonian is

$$\hat{H}_{lep} = -\frac{1}{2m_p}\nabla_1^2 - \tfrac{1}{2}\nabla_2^2 - \tfrac{1}{2}\nabla_3^2 + \frac{1}{r_{A1}} + \frac{1}{r_{B1}} - \frac{1}{r_{A2}} - \frac{1}{r_{B2}}$$
$$- \frac{1}{r_{A3}} - \frac{1}{r_{A3}} + \frac{1}{R} - \frac{1}{r_{12}} - \frac{1}{r_{13}} + \frac{1}{r_{23}}, \tag{14}$$

where A and B are the nuclei. The potential is made up of the Coulomb interactions between all five particles. The nuclei are fixed at internuclear distance R. The leptonic coordinates are fixed relative to the nuclei. For details of the relationship between molecule fixed coordinates and non-rotating coordinates, see, for example, Armour and Jonsell [15].

The Schrödinger equation for the leptons is

$$\hat{H}_{\text{lep}}\Psi_{\text{lep},m}(\boldsymbol{r}_1, \boldsymbol{r}_2, \boldsymbol{r}_3; \boldsymbol{R}) = E_m(R)\Psi_{\text{lep},m}(\boldsymbol{r}_1, \boldsymbol{r}_2, \boldsymbol{r}_3; \boldsymbol{R}) \tag{15}$$

In principle, this equation can be solved exactly. Suppose a weakly bound state is found for a range of R values surrounding the equilibrium R value of H_2.

There are two possibilities.

(1) The state is bound for all R values for which the associated vibrational wave function has an appreciable non-zero value.

(2) The state is only bound for a subset of the above R values.

We will only consider possibility (1). Denote the bound state by $m = 0$. There will also be continuum solutions to the Schrödinger (15). Denote them by $m = 1$. They will have energy

$$E_1(R) = E_p + E_{mol}(R), \tag{16}$$

where E_p is the energy of the incident positron and $E_{mol}(R)$ is the electronic energy of H_2 at a given R value. In our earlier calculations, we obtained approximate solutions of this type at a given R value, in most cases the H_2 equilibrium value $1.4a_0$.

In the B–O approximation, we obtain normalised bound states

$$\{\Psi_{lep,0}(\boldsymbol{r}_1, \boldsymbol{r}_2, \boldsymbol{r}_3; \boldsymbol{R})\chi_{0n}(\boldsymbol{R})\}, \quad n = 0, 1, 2, \ldots$$

where $\chi_{0n}(\boldsymbol{R})$ is a vibrational wave function associated with the potential $E_0(\boldsymbol{R})$, with appropriate angular momentum.

The bound state with $n = 0$, i.e. corresponding to the vibrational ground state, will be below the continuum. But when $n > 0$, the states will be in the continuum of states $\{\Psi_{\text{lep},1}(\boldsymbol{r}_1, \boldsymbol{r}_2, \boldsymbol{r}_3; \boldsymbol{R})\chi_{10}(\boldsymbol{R})\}$. However, no interaction with the continuum will occur if we remain within the B–O approximation. Coupling will only occur if we go beyond the B–O approximation. These bound states will then become quasi-bound.

In Gribakin's theory [16], it is the existence of a permanent molecular dipole moment that brings about coupling between the quasi-bound states and the continuum. H_2 does not have a permanent dipole moment. However, such a dipole moment could be introduced into our model by changing the unit charges on the nuclei to $1+Z$ and $1-Z$, where $0<Z<1$. We have not thought it necessary to do this as the coupling between the quasi-bound states and the continuum is due to the breakdown of the B–O approximation. Further investigation is necessary to find out if any relationship between this and a permanent dipole moment can be derived.

We have carried out a preliminary Kohn calculation to investigate the effect on $Z_{\rm eff}(k)$ of the coupling between the lowest quasi-bound state, $\Psi_{{\rm lep},0}\chi_{01}$ with zero angular momentum and the continuum due to the breakdown of the B–O approximation. We take the target H_2 to be in its ground state with zero total angular momentum. We find analytically that resonant behaviour occurs when the energy E of e^+H_2 satisfies the relation

$$E = E_{01} + B, \tag{17}$$

where E_{01} is the energy of the quasi-bound state, as calculated using the B–O approximation. B is a level shift equal to the first order correction to E_{01} due to the inclusion of the terms neglected in the B–O approximation.

If we use a basis set containing $\Psi_{{\rm lep},0}\chi_{01}$ with all other basis functions orthogonal to it, we find from the Kohn equations that the coefficient of $\Psi_{{\rm lep},0}\chi_{01}$ in the scattering wave function is of $O(\mu_M)$, when E satisfies (17). The contribution to $Z_{\rm eff}(k)$ at this E value from this quasi-bound state is thus of $O(\mu_M^2)$. As $\mu_M = 918$, at this E value $Z_{\rm eff}(k)$ is likely to have a large value.

This theory should be extendable to organic molecules that have weakly bound positron-molecule states associated with very large values of $Z_{\rm eff}(k)$.

2.3. *Examination of singular behaviour in generalised Kohn calculations*

Running parallel to our calculations of $Z_{\rm eff}(k)$, we have carried out an examination of singularities that can arise when carrying out calculations using the generalised Kohn method in which a parameter τ is added to $c(\lambda_1 - 1)$ in the arguments of the sine and cosine factors in equations (5) and (6) and T in equation (4) is taken to be equal to the modified C factor. A detailed study is made of the structure of the Kohn equations and the behaviour of the phase shift as τ is varied. A method is developed for determining an optimum choice of τ to avoid anomalous behaviour due to the singularities. This optimum choice is shown to give a phase shift value that is the same as that obtained by taking the real part of the phase shift calculated using the complex Kohn method [10, 17, 18].

3. Helium-antihydrogen scattering

The calculations that we have carried out on He$\bar{\rm H}$ scattering are a continuation of those described in [4]. We are interested in the rearrangement reactions

$$\mathrm{He} + \bar{\mathrm{H}} \rightarrow \mathrm{He}^+\bar{p} + \mathrm{Ps} \tag{18}$$

$$\mathrm{He} + \bar{\mathrm{H}} \rightarrow \mathrm{He}\bar{p} + e^+ \tag{19}$$

$$\mathrm{He} + \bar{\mathrm{H}} \rightarrow \alpha\bar{p} + \mathrm{Ps}^- \tag{20}$$

We calculate the entrance channel wave function using the B–O approximation and the Rayleigh-Ritz variational method. The coordinates we use are prolate spheroidal coordinates, as in the e^+H_2 calculation. We take $\bar{p}$ to be at A and α at B in Figure 1. The basis functions we use to calculate the leptonic part of the entrance wave function at a given R are of the form

$$\psi_i = S_{23} N \lambda_1^{a_i} \lambda_2^{b_i} \lambda_3^{c_i} \mu_1^{d_i} \mu_2^{e_i} \mu_3^{f_i} C(s_i, t_i) \times \exp[-\alpha_1\lambda_1 - \alpha_2\lambda_2 - \alpha_3\lambda_3 + \beta_1\mu_1 + \beta_2\mu_2 + \beta_3\mu_3], \quad (21)$$

where S_{23} is the electron symmetriser and N is a normalisation constant. $a_i, b_i, c_i, d_i, e_i, f_i$ are non-negative integers and $\alpha_1, \alpha_2, \alpha_3, \beta_1, \beta_2, \beta_3$ are non-linear parameters. The electrons are in a singlet spin state. These basis functions are the same as those used in the short-range correlation functions in the e^+H_2 scattering calculations but with μ_1, μ_2 and μ_3 in the exponent, in addition to λ_1, λ_2 and λ_3. This is to take into account the asymmetry of the nuclei. The factor $C(s_i, t_i)$ is as given in Table 1.

The close relationship between these basis functions and those used in the e^+H_2 calculations made it possible to use the same methods to calculate the required matrix elements. In fact, the methods were first developed in the context of the He$\bar{\text{H}}$ calculations. Details of the variational calculations of the He$\bar{\text{H}}$ wave functions at a given R value and the associated energies are given in [4].

Cross sections for reactions (18) and (19) and preliminary results for (20) are given in [4]. We wish to describe further calculations that we have carried out for reaction (20). We would like eventually to carry out calculations on these reactions using the Kohn variational method, as in the case of the H$\bar{\text{H}}$ rearrangement reaction [3]:

$$\text{H} + \bar{\text{H}} \rightarrow \text{Pn} + \text{Ps}, \quad (22)$$

where Pn is protonium ($p\bar{p}$). However, such a calculation would be very complicated to carry out. For this reason, we have used a method involving the T-matrix and a distorted wave approximation.

The rearrangement cross section, σ_{3i} for the third He$\bar{\text{H}}$ reaction, i.e. reaction (20), is of the form

$$\sigma_{3i} = \frac{(2\pi)^4}{\hbar v_i} \int \delta(E_i - E_3) |T_{3i}|^2 d\boldsymbol{k}_3, \quad (23)$$

where v_i is the relative speed of the colliding particles He and $\bar{\text{H}}$, $\boldsymbol{k}_3$ is the wave vector of the relative motion of $\alpha\bar{p}$ and Ps^- and E_i and E_3 are the total energies of the initial and final states, respectively. T_{3i} is the T-matrix element

$$T_{3i} = \langle \Phi_3 | V_3 | \Psi_{\boldsymbol{k}_i}^{(+)} \rangle, \quad (24)$$

where $\Psi_{\boldsymbol{k}_i}^{(+)}$ is the exact scattering wave function for the incident energy under consideration and $\boldsymbol{k}_i$ is the associated wave vector. V_3 is the potential that couples the two systems resulting from the scattering, i.e. $\alpha\bar{p}$ and Ps^-. Φ_3 is the appropriate wave function to represent $\alpha\bar{p}$ and Ps^- in the rearrangement channel.

It is not possible to determine $\Psi_{\boldsymbol{k}_i}^{(+)}$. Thus we used the form of the distorted wave approximation introduced by Jonsell *et al* [19] in their calculation of rearrangement cross sections for H$\bar{\text{H}}$ scattering. In this method, $\Psi_{\boldsymbol{k}_i}^{(+)}$ is approximated by the wave function for the entrance channel, calculated using the Born-Oppenheimer approximation.

The leptonic wave function for the incident channel, $\psi_{\rm BO}(\boldsymbol{r}; R)$, was calculated using the basis functions (21) and the variational method as described in [4]. $\boldsymbol{r}$ represents the coordinates of the leptons and $\boldsymbol{R}$ is the internuclear vector. The continuum wave functions, $\chi(\boldsymbol{R})$, for the relative motion of He and $\bar{\rm H}$ was calculated using the very accurate B–O potential of Strasburger and Chojnacki [20]. The approximate wave function we used in place of the exact wave function $\Psi^{(+)}_{\boldsymbol{k}_i}$ in (24) is $\psi_{\rm BO}(\boldsymbol{r}; R)\chi(\boldsymbol{R})$.

The wave function, Φ_3, used to represent the rearrangement channel was of the form

$$\Phi_3 = \eta(\boldsymbol{\rho})\phi_{Ps^-}(r_{12}, r_{13}, r_{23})\chi_{\alpha\bar{p}}(\boldsymbol{R}) \tag{25}$$

where $\chi_{\alpha\bar{p}}(\boldsymbol{R})$ is an $\alpha\bar{p}$ bound state wave function. This is taken to be an s-state. $\phi_{\rm Ps^-}(r_{12}, r_{13}, r_{23})$ was taken to be the 15 term ground-state $\rm Ps^-$ wave function calculated by Ho [21]. The energy value of this ground state obtained using this wave function is -0.26161. The exact value is -0.26201 [22]. $\eta(\boldsymbol{\rho})$ is the wave function for the $\rm Ps^-$ centre of mass motion, where $\boldsymbol{\rho}$ is the position vector of the centre of mass of $\rm Ps^-$ w.r.t. the centre of mass of $\alpha\bar{p}$. If $\eta(\boldsymbol{\rho})$ is an s-wave, it is a function of ρ only. For higher partial waves, it will depend on the spherical polar angles θ_ρ and ϕ_ρ of $\boldsymbol{\rho}$. These angles must be in terms of fixed-axes that do not rotate with $\boldsymbol{R}$. $\eta(\boldsymbol{\rho})$ can be expressed in this way using the rotation properties of spherical harmonics.

The T-matrix element (24) in the distorted wave approximation can be expressed in the form

$$T_{3i} = \int \chi^*_{\alpha\bar{p}}(\boldsymbol{R})t_{3i}(R)\chi(\boldsymbol{R})\,d\boldsymbol{R} \tag{26}$$

where

$$t_{3i}(R) = \langle\eta(\boldsymbol{\rho})\phi_{Ps^-}(r_{12}, r_{13}, r_{23})|V_3|\psi_{\rm BO}(\boldsymbol{r}; R)\rangle \tag{27}$$

is the leptonic T-matrix. The integration in the expression for $t_{3i}(R)$ is over the leptonic coordinates. $t_{3i}(R)$ was evaluated using 8D numerical integration, rather than 9D, as the symmetry about the internuclear axis made it possible to reduce the dimension of the integrals by one.

An interesting question arises regarding the form of V_3 in the T-matrix, T_{3i}. The potential between $\alpha\bar{p}$ and $\rm Ps^-$ is, in an obvious notation,

$$V = \frac{2}{r_{\alpha 1}} - \frac{2}{r_{\alpha 2}} - \frac{2}{r_{\alpha 3}} - \frac{1}{r_{\bar{p}1}} + \frac{1}{r_{\bar{p}2}} + \frac{1}{r_{\bar{p}3}}. \tag{28}$$

As

$$V \sim_{\rho\to\infty} -\frac{1}{\rho}, \tag{29}$$

it is of attractive Coulumb form asymptotically. As is well-known, such a potential affects the form of the scattering wave function, $\eta(\boldsymbol{\rho})$, asymptotically.

We are in the process of considering the following forms for V_3 and $\eta(\boldsymbol{\rho})$

(1) $V_3 = V$ and $\eta(\boldsymbol{\rho}) = F_c(\boldsymbol{\rho})$, where $F_c(\boldsymbol{\rho})$ is the Coulomb wave function in a potential $-\dfrac{1}{\rho}$

(2) $V_3 = V + \dfrac{1}{\rho}$ and $\eta(\boldsymbol{\rho}) = F_c(\boldsymbol{\rho})$

(3) $V_3 = V + \dfrac{1}{\rho} + \dfrac{\mu}{k^2(1 + \cos w)\rho^2}$,

where μ is the reduced mass of $\alpha\bar{p}$ and Ps^-, k is the wave number of the relative motion and w is the angle between the directions of $\boldsymbol{k}$ and $\boldsymbol{\rho}$.

$$\eta(\boldsymbol{\rho}) = \exp\left\{i\left[\boldsymbol{k}\cdot\boldsymbol{\rho} + \frac{\mu}{k}\ln(k\rho + \boldsymbol{k}\cdot\boldsymbol{\rho})\right]\right\}.$$

This is a method of taking into account the Coulomb form of V asymptotically that we are currently investigating. It is related to the methods developed by Mulherin and Zinnes [23], Rosenberg [24] and Belkić *et al* [25].

Another possible way of treating the Coulomb form of V that we are considering is the surface-integral formulation of scattering theory of Kadyrov *et al* [26].

We have obtained rearrangement cross sections for s-wave scattering using the first form of V_3 and $\eta(\boldsymbol{\rho})$ above. This is not a very satisfactory choice from a theoretical point of view. However, it is the easiest to use and should give a reasonable estimate of the size of the cross sections. We estimate that higher partial waves must be included for energies $>\sim 10^{-6}$ a.u.

The results obtained for the cross sections are given in Table 3. An error in the calculation described in the second paper in [4] has been corrected. The threshold energy is between $\alpha\bar{p}$ states with principal quantum number $N_p = 30$ and $N_p = 31$.

In the range considered, the cross sections $\sigma_{3i}(N_p)$ only differ slightly from their threshold behaviour given by

$$\lim_{\varepsilon_i \to 0+}\left[\varepsilon_i^{\frac{1}{2}}\sigma_{3i}(N_p)\right] = C_N \tag{30}$$

where ε_i is the incident energy and C_N is a constant that depends on the $\alpha\bar{p}$ principal quantum number, N_p. Both $\chi_{\alpha\bar{p}}(R)$, for the high N_p values under consideration, and $\chi(R)$ are rapidly oscillating functions of R. (As they are s-states, they are only functions of R). The leptonic T-matrix elements do not depend on N_p and there is evidence that they vary smoothly with R [27]. Thus, as would be expected from (26) and (27), we find some structure in $\sigma_{3i}(N_p)$, depending on the extent to which the oscillation in $\chi_{\alpha\bar{p}}(R)$ and $\chi(R)$ are in phase at a given incident energy. As this structure is very sensitive to the values of these wave functions, we wish to do further checks before publishing graphs of $\sigma_{3i}(N_p)$.

It can be seen that all the cross sections in Table 3 are less than or equal to $\sim 10^{-3}\varepsilon_i^{-\frac{1}{2}}a_0^2$. The antiproton interacts with a proton, or a nucleus containing a proton, through the strong interaction to form pions. The strong interaction is much stronger than the electromagnetic interaction that brings about the annihilation of a positron and an electron. At very low energies the cross section for antiproton loss through the strong interaction in He$\bar{H}$ scattering is $\sim 2.6\varepsilon_i^{-\frac{1}{2}}a_0^2$, which is very much larger than the cross sections in Table 3. It is also much larger than the cross sections for the other two rearrangement processes (18) and (19). See [4].

4. Conclusion

We have carried out detailed calculations on two scattering processes, low-energy positron-hydrogen-molecule and helium-antihydrogen scattering that contain the same three leptons.

In the calculations of e^+H_2 scattering, we used the complex Kohn variational method to calculate $Z_{\text{eff}}(k)$. We were able to avoid using the method of models [5] that was a feature of previous calculations [6]. This was possible as we were able to calculate integrals involving $\frac{\rho_{12}\rho_{13}}{\rho_{23}}$ [9]. The results of our calculations are given in [8].

Recently, we have examined the behaviour of $Z_{\rm eff}(k)$ when the internuclear distance, R, is varied. When the nuclear motion is taken into account, we obtain a value of $Z_{\rm eff}(k)$ considerably closer to experiment using the method of models. This is partly due to the much larger values of $Z_{\rm eff}(k)$ that are obtained at $R > 1.6\,a_0$ using the method of models. Similar behaviour at $R > 1.6a_0$ has very recently been found by Zhang *et al* [11] in a very accurate calculation.

We have also used a model e^+H_2 system and the Kohn method to investigate possible resonant behaviour due to the presence of a quasi-bound positron-H_2 state, associated with an excited vibrational state. A preliminary calculation shows that such a state is coupled to the continuum through a breakdown in the Born-Oppenheimer approximation, probably resulting in a large $Z_{\rm eff}(k)$ value when a resonant condition is satisfied. This is an interesting result. It is not clear how it is related to Gribakin's theory [13, 16] of the very large $Z_{\rm eff}(k)$ values that have been observed in positron scattering by some organic molecules [14].

We have also carried out interesting studies of singular behaviour in the generalised Kohn method and also of the relationship between phase shifts obtained by this method and the complex Kohn method [10, 17, 18].

In previous calculations on low-energy $\rm H\bar{H}$ scattering [4], we carried out calculations cross sections for the rearrangement reactions (18) and (19). Preliminary calculations of cross sections were also carried out for the rearrangement reaction (20). These made use of a 15-term $\rm Ps^-$ ground-state wave function obtained by Ho [21].

We have carried out further calculations for the third reaction and corrected an error in the preliminary calculations [4]. The potential between $\alpha\bar{p}$ and $\rm Ps^-$ is of Coulomb form asymptotically. We are in the process of investigating various possible ways of taking this into account [22–25].

We obtained values for the s-wave contribution to the cross section for the first four $\alpha\bar{p}$ s-states below threshold, using the simplest method of treating the long-range Coulomb behaviour. Contributions from higher partial waves must be included for incident energies greater than about 10^{-6} a.u., which is in the sub-Kelvin temperature range.

In the calculations that we have carried out for reaction (20), and in [4] for reactions (18) and (19), we find that the rearrangement cross sections that we obtain are much smaller than the antiproton annihilation cross-section.

Acknowledgments

We are grateful to John Humberston for valuable discussions, Mike Charlton for encouraging us to work on positron and antimatter physics and to Eugene Ho for sending us his $\rm Ps^-$ ground-state wave function. We wish to thank EPSRC (UK) for support for this research through an advanced fellowship grant (EP/D069785/1), two doctoral training grants, (GR/P01595/01) and (EP/P500389/1), and a project studentship (EP/C548019/1).

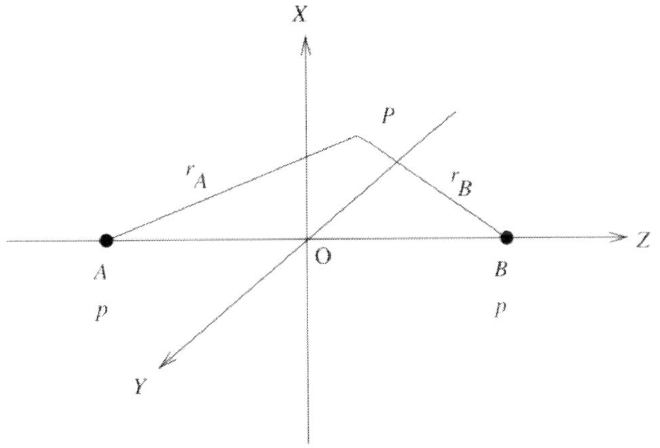

Figure 1. Coordinate System

Table 1. Forms used for the factor $C(s_i, t_i)$ in the short range correlation functions.

s_i	t_i	$C(s_i, t_i)$	Type of basis function
0	0	1	Product of σ functions
1	1	ρ_{12}	Positron-electron Hylleraas-type
1	2	ρ_{23}	Electron-electron Hylleraas-type
2	1	$\frac{2}{R}(x_1x_2 + y_1y_2)$	Positron-electron configuration interaction (CI) product of π functions
2	2	$\frac{2}{R}(x_2x_3 + y_2y_3)$	Electron-electron CI, product of π functions

R is the internuclear distance.
The distance between particles i and j,

$$r_{ij} = \frac{R}{2}\rho_{ij}.$$

Table 2. The dependence of $Z_{\text{eff}}(k)$ for $k = 0.04$ on the internuclear distance R.

Internuclear distance, R (a_0)	$Z_{\text{eff}}(k)$	
	Calculated using the method of models †	Calculated without using the method of models *
0.8	6.4	5.0
0.9	7.2	5.8
1.0	8.0	6.5
1.2	10.0	8.1
1.4	12.6	9.6
1.6	15.8	10.6
1.8	19.8	11.1
2.0	24.3	10.8
2.1	26.7	10.5

†Trial function containing 279 short-range correlation functions, 18 of which correlate the positron and the electrons, and a target wave function that takes into account 57.1% of the correlation of energy of H_2 at $R = 1.4a_0$.
*Trial function as above but with a target wave function that takes into account 96.8% of the correlation energy of H_2 at $R = 1.4a_0$.

Table 3. Cross sections for rearrangement into $\alpha\bar{p} + \text{Ps}^-$.

Incident energy ε_i (a.u.)	Cross sections $(\sigma_{3i}(N))$ $(\varepsilon_i^{-\frac{1}{2}} a_0^2)$			
	$N = 30$	$N = 29$	$N = 28$	$N = 27$
10^{-10}	0.36×10^{-3}	0.30×10^{-3}	0.004×10^{-3}	0.017×10^{-3}
10^{-8}	0.36×10^{-3}	0.30×10^{-3}	0.004×10^{-3}	0.017×10^{-3}
10^{-6}	0.31×10^{-3}	0.26×10^{-3}	0.004×10^{-3}	0.015×10^{-3}

N is the principal quantum number of the s-state $\alpha\bar{p}$.

References

[1] Fujiwara M *et al.* 2008 Phys. Rev. Lett. **101**, 053401.
[2] Gabrielse G *et al.* 2008 Phys. Rev. Lett. **100**, 113001.
[3] Armour E A G and Chamberlain C W, 2002 J. Phys. B: At. Mol. Opt. Phys. **35** L489.
[4] Armour E A G, Todd A C, Jonsell S, Liu Y, Gregory M R and Plummer M, 2008 Nucl. Instr. and Meth. B **266**, 363; in 'Cold Antimatter Plasmas and Application to Fundamental Physics', 2008 (eds. Kanai Y and Yamazaki Y) AIP CP1037, 333.
[5] Drachman R J, 1972 J. Phys. B: At. Mol. Opt. Phys. 5 L30.
[6] Armour E A G, Baker D J and Plummer M, 1990 J. Phys. B: At. Mol. Phys. **23**, 3057.
[7] Laricchia G, Charlton M, Beling C D and Griffith T C, 1987 J. Phys. B: At. Mol. Phys. **20**, 1865.
[8] Cooper J N, Armour E A G and Plummer M, 2008 J. Phys. B: At. Mol. Opt. Phys. **41**, 245201.
[9] Plummer M, Armour E A G, Todd A C, Franklin C P and Cooper J N, 2009 Computer Physics Communications, in press.
[10] Cooper J N 2009 PhD thesis, University of Nottingham, UK.
[11] Zhang J-Y, Mitroy J and Varga K, 2009 Phys. Rev. Lett. **103**, 223202.
[12] Van Reeth P and Humberston J W, 1995 J. Phys. B: At. Mol. Opt. Phys. **28**, L23.
[13] Gribakin G F, 2000 Phys, Rev, A **61**, 022720.
[14] Gilbert S J, Barnes L D, Sullivan J P and Surko C M, 2002 Phys. Rev. Lett. **88**, 043201
[15] Armour E A G and Jonsell S, 2007 J. Phys. A: Math. Theor. **40**, 701.
[16] Gribakin G F and Lee C M R, 2006 Phys. Rev. Lett. **97**, 193201
[17] Cooper J N, Armour E A G and Plummer M, 2009 J. Phys. A: Math. Theor. **42**, 095207.
[18] Cooper J N, Plummer M and Armour E A G, to be published.
[19] Jonsell S, Saenz A, Froelich P, Zygelman B and Dalgarno A, 2004 J. Phys. B: At. Mol. Opt. Phys. **37**, 1195.
[20] Strasburger K and Chojnacki H, 2002 Phys. Rev. Lett. **88**, 163201.
[21] Ho Y K, 2007 Private communication.
[22] Ho Y K, 1993 Phys. Rev. A **48** 4780.
[23] Mulherin D and Zinnes I I, 1970 J. Math. Phys. **11**, 1402.
[24] Rosenberg L, 1972 Phys. Rev. D **8**, 1833.
[25] Balkić D Ž, Gayet R and Salin A, 1979 Phys. Rep. **56**, 279.
[26] Kadyrov A S, Bray I, Mukhamedzhanov A M and Stelbovics A T, 2005 Phys. Rev. A **72**, 032712; 2009 Annals of Physics **324**, 1516.
[27] Gregory M R, 2007 PhD thesis, University of Nottingham, UK.

Journal of Physics: Conference Series **199** (2010) 012008 doi:10.1088/1742-6596/199/1/012008

Simulations of antihydrogen formation in a nested Penning trap

S. Jonsell, D. P. van der Werf and M. Charlton

Department of Physics, Swansea University, Swansea SA2 8PP, United Kingdom

E-mail: b.s.jonsell@swansea.ac.uk

Abstract. We have simulated the formation of antihydrogen through three-body recombination using classical trajectories of antiprotons and positrons. The simulations include several effects which are important in current antihydrogen experiments: the full motion of the antiproton repeatedly passing into and out of the positron plasma, the energy loss of antiprotons due to the interaction with the positron plasma, and the field-ionization of antihydrogen en route from the plasma to the detector. We find that whereas the overall simulated rate of formation of antihydrogen has a density dependence close to n_e^2, the rate of antihydrogen detection follows a power law less than 2. The difference is due to the effect of density dependent field ionization.

1. Introduction

Antihydrogen can be used for precision tests of fundamental matter-antimatter symmetries. This is the goal of a number of experiments using the Antiproton Decelerator (AD) at CERN. Here the first experiments, ATHENA [1] and ATRAP [2], managed to form cold antihydrogen in 2002. The efforts to trap and cool antihydrogen continue. Presently, two experiments, ALPHA [3] and ATRAP [4], are working towards the goal of trapping antihydrogen for future spectroscopic tests of the CPT symmetry. There are also new experiments underway from the ASACUSA and AEGIS collaborations [5]. Along with the experimental efforts there has been a lot of theoretical work trying to interpret results and suggest new strategies. For a recent review see Ref. [6]. A good understanding of the formation process is necessary to guide the experiments. This is the motivation for the simulations of antihydrogen formation presented in this paper.

In the ATHENA, ALPHA and ATRAP experiments antihydrogen is formed from antiprotons and positrons trapped together in a nested Penning trap. At the relevant temperature and density the most prominent formation mechanism is three-body collisions,

$$\bar{\mathrm{p}} + e^+ + e^+ \to \bar{\mathrm{H}} + e^+. \tag{1}$$

Through this process weakly bound Rydberg states of antihydrogen are created. To stabilize the anti-atom its binding energy has to be increased through further collisions. However, more often the anti-atom will be ionized again, either through collisions with positrons or by the electric field present in the trap. An alternative formation mechanism is the radiative process,

$$\bar{\mathrm{p}} + e^+ \to \bar{\mathrm{H}} + \gamma. \tag{2}$$

This process will yield more tightly bound antihydrogen, and is therefore interesting even though it is likely to be much more rare. In our present simulations the radiative process is not included.

Table 1. Radii and lengths of positrons plasmas for different plasma densities n_e and particle numbers N_e. In all cases $T_e = 15$ K.

n_e (m^{-3})	N_e	radius (mm)	length (mm)
5×10^{13}	1.4×10^7	2.87	16.52
1×10^{14}	2.8×10^7	2.41	23.56
2×10^{14}	5.6×10^7	2.06	32.16
5×10^{14}	1.4×10^8	1.70	51.18
1×10^{15}	1.2×10^8	1.11	51.75

2. Method

We simulate the experiments by calculating trajectories of antiprotons in the nested Penning trap. The simulations start with the initialization of antiprotons moving along the axis of the trap with some pre-set kinetic energy, in most cases 2 eV. The path of the antiprotons back and forth though the positron plasma is then followed, with the probability of collision with a positron calculated in each time step. A collision is defined as an event when a positron comes within a box with side length $n_e^{-1/3}$ (where n_e is the density of the positron plasma) centered at the antiproton. As long as the positron stays inside this box its trajectory is calculated, along with that of the antiproton and those of any positron already present or later entering the box. The operational definition of an antihydrogen atom is a state with one positron inside this box. The interaction between the antiproton and positrons outside the box is included as an energy loss of a charged particle in an oppositely charged and magnetized medium [7]. The simulation ends when an antiproton in the form of antihydrogen reaches the detectors surrounding the trap or when it is trapped outside the positron plasma and therefore lost for antihydrogen formation. Such trapping of antiprotons is the result of field ionization of loosely bound antihydrogen by the electric fields in the region between the positron plasma and the detectors. More details can be found in [8].

The calculations are purely classical, using an adaptive step-size Runge-Kutta algorithm to integrate Newton's equation of motion. The force on the antiproton and any positrons present include their mutual Coulomb interaction, along with the force from the external electric and magnetic fields. The magnetic field is constant and directed along the axis of the trap, thus providing the radial confinement of the particles. In most simulations, we set the magnetic field to $B = 3$ T, which is the value used by the ATHENA experiment [1]. The electric field arises from the cylindrical electrodes surrounding the trap and from the space charge of the positron plasma. The electric field and the density distribution of the positrons are determined by solving Poisson's equation self-consistently at the appropriate positron temperature, in the simulations presented here set to $T_e = 15$ K [9]. Other input parameters to the calculation of the electric fields are the peak density of the positron plasma and the number of positrons. We used plasma densities between 5×10^{13} m^{-3} and 10^{15} m^{-3} and positron numbers between 1.4×10^7 and 1.4×10^8. This corresponds typical experimental parameters, for instance in [1] the plasma density was 2.5×10^{14} m^{-3} and the positron number 7×10^7. The plasma parameters are summarised in Table 1. The number of antiprotons in the experiments, typically a few 1000 per mixing experiment, is small enough to justify the neglect of antiproton-antiproton interactions.

From our simulations we can extract distributions of velocities, binding energies etc. of the detected antihydrogen. We can also extract spatial and time distributions of antihydrogen formation. Moreover, we can improve our understanding of the formation process by extracting properties of the antiprotons/antihydrogen at intermediate times before their detection.

Journal of Physics: Conference Series **199** (2010) 012008 doi:10.1088/1742-6596/199/1/012008

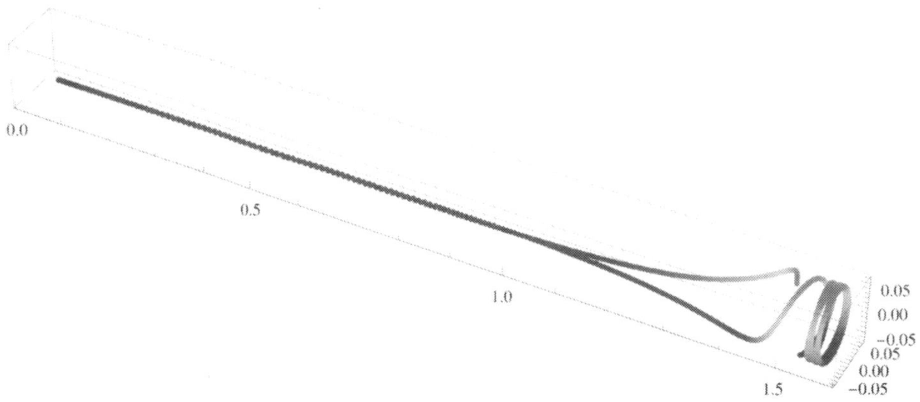

Figure 1. Part of an antiproton trajectory in a nested Penning trap. Inside the positron plasma (to the right) the antiproton follows a helical magnetron motion around the axis of the trap (the radius of the cyclotron motion is too small to be resolved). To the left the antiproton leaves the plasma, is reflected back by a barrier in the electric potential and returns to the positron region. The length scales on the axes are in cm.

3. Formation rates

Antihydrogen formation is a multi-step process which requires time. The final states of the three-body process in eq. (1) are too loosely bound to have any chance to survive the electric fields between the plasma and the detector. The antihydrogen must therefore stabilize by gaining binding energy through repeated collisions with positrons. Such collisions may also have the opposite effect, namely reducing the binding energy or ionizing the antihydrogen. Therefore only a small fraction of the antihydrogen atoms initially formed will eventually leave the plasma, either to be detected or field ionized outside the positron plasma.

The rate of recombination is often described as the rate at which atoms pass a "bottleneck" at a binding energy of a few $k_{\mathrm{B}}T_{\mathrm{e}}$. Atoms with binding energies greater than this bottleneck are very unlikely to be re-ionized. For an antiproton immersed in a plasma with temperature T_{e} and density n_{e}, the formation rate has been calculated to be $C(e^2/(4\pi\epsilon_0))^5 m_{\mathrm{e}}^{-1/2} n_{\mathrm{e}}^2 (k_{\mathrm{B}}T_{\mathrm{e}})^{-9/2}$ (here m_{e} is the electron mass and the other constants have their usual meaning), where $C = 0.76$ for $B = 0$ [10], $C = 0.11$ for $B = 3$ T [11] and $C = 0.070$ for $B = \infty$ [12].

These rates were derived assuming that the antihydrogen with binding energies above the bottleneck reach a steady state distribution, from which anti-atoms trickle down towards the ground state. This is, however, not a good description of the current antihydrogen experiments. As was pointed out by Robicheaux [13], an important feature of these experiments is that the antiprotons only spend a short time inside the positron plasma. The antiprotons traverse the positron plasma along the axis of the trap. After leaving the plasma they pass a well in the electric potential energy (which for the oppositely charged positrons is a barrier providing their axial confinement, and thus a necessary feature of the experiment), before they are reflected back by a barrier in the electric potential. (An example of part of an antiproton trajectory is shown in Figure 1.) Hence, there will not be time to establish a steady-state distribution, and indeed very few antihydrogen have binding energies greater than the bottleneck energy. Instead, the distribution of binding energies will be a snap-shot taken at the time the antihydrogen leaves the plasma. Most of the antihydrogen will therefore be very loosely bound, and a large fraction will not survive the electric fields between the positron plasma and the detector. Field ionization will usually lead to trapping of the antiprotons outside the positron plasma, and is hence a loss process for antiprotons.

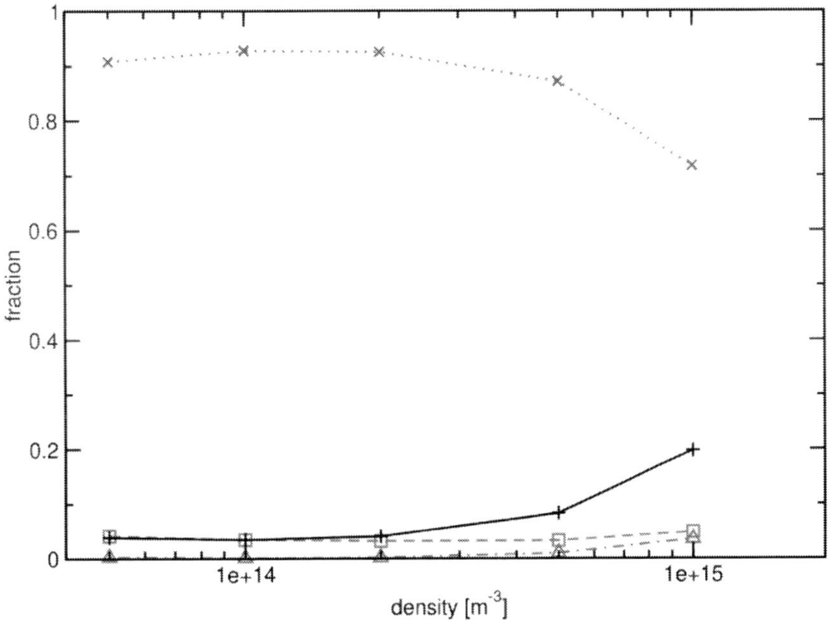

Figure 2. Fraction of antihydrogen atoms formed with a binding energy of at least k_BT_e that re-ionize within the plasma (blue, ······, ×), are field ionized outside the plasma leading to loss of the antiproton (black, ——, +), survive to the detector (red, - - - -, □) or ionized with the antiproton returning to the plasma (green, — · —, △). The magnetic field was $B = 3$ T and the plasma temperature $T_e = 15$ K.

The relative probabilities of the various possible outcomes for an antihydrogen atom formed in the plasma are shown in Figure 2 as a function of plasma density. Here an antihydrogen atom with kinetic energy from a thermal distribution with temperature $T_e = 15$ K and binding energy greater than k_BT_e was formed at a random position inside the positron plasma. The status of the antiproton as it left the plasma (either as a single particle or inside an antihydrogen atom) was recorded. We find that around 90% of the time the antihydrogen is ionized before it leaves the plasma for most densities. The probability that the antihydrogen survives to the detector is relatively constant, around 3–4% for all densities. The drop in the re-ionization probability at the highest density arises because of repeated formation events in a single pass through the plasma. In fact, at $n_e = 10^{15}$ m^{-3} the antiproton on average goes through 5 cycles of antihydrogen formation and re-ionization in a single pass through the plasma.

The effect of field ionization is evident in Figure 3. Here we compare the distribution of binding energies of antihydrogen atoms just after they have left the plasma to the same distribution for antihydrogen reaching the detector. We see that almost all antihydrogen with binding energies less than 50 K, and the majority of those with binding energy between 50 K and 80 K, have been field ionized. This constitutes the majority, about 82%, of the antihydrogen formed. This fraction is density dependent, with a smaller fraction of field-ionized antihydrogen at lower positron densities. Hence, the rate of antihydrogen detection is modified in a density dependent way. This is shown in Figure 4. Here we find that the total formation rate scales with density to a power just below 2, whereas the rate of antihydrogen detection is significantly reduced, scaling as $n_e^{1.67}$.

We are presently working on the temperature dependence of the formation rate. Preliminary

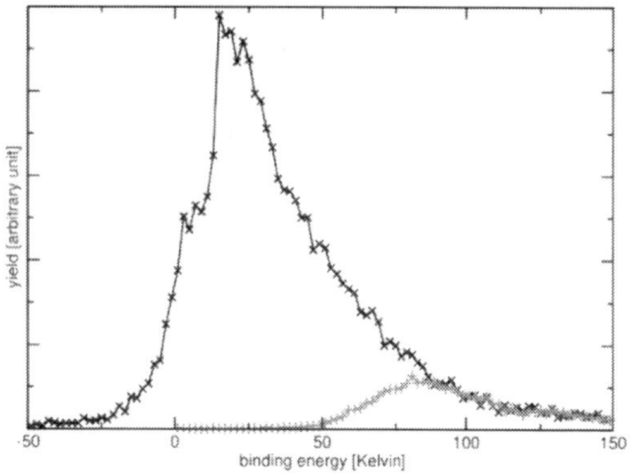

Figure 3. Distribution of binding energies for antihydrogen just after leaving the positron plasma (black, ——, ×) and when reaching the detector (red, - - - -, +). 18% of the antihydrogens reach the detector. In this simulation $n_e = 10^{15}$ m^{-3}, T_e = 15 K, and B = 3 T. The small number of antihydrogens with negative binding energies are metastable and dissociate before they reach the detector.

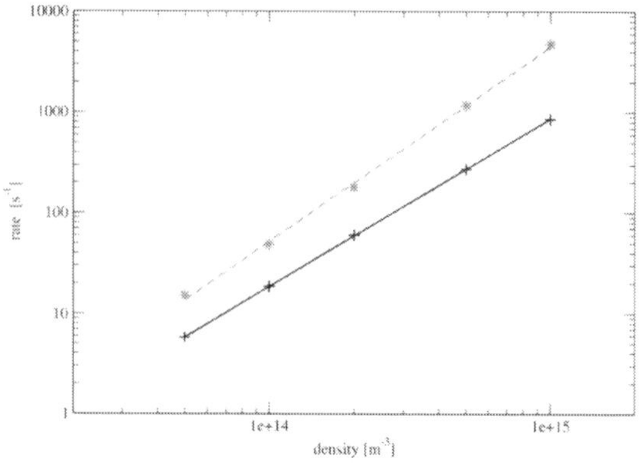

Figure 4. Antihydrogen formation rates as a function of positron density: (black, ——, +) rate of antihydrogen detection, (red, - - - -, $*$) total formation rate (including antihydrogen that is field ionized outside the plasma). In this simulation T_e = 15 K, and B = 3 T. The solid lines are power-law fits, giving the exponents 1.67 ± 0.01 for detected antihydrogen and 1.93 ± 0.05 for the total rate.

results indicate that the temperature scaling is less steep than the $T_e^{-9/2}$ predicted for the steady-state situation. This is qualitatively in accordance with recent experiments [14].

4. Conclusions

We have studied the dependence of antihydrogen formation rates from three-body recombination on the density of the positron plasma. We find that the short time the antiproton spends in the positron plasma, as pointed out in Ref. [13], along with the field-ionization of loosely bound antihydrogen, has a profound effect on the formation rate. In particular, the antihydrogen detection rate has a weaker density dependence than the n_e^2 one would normally expect for a three-body process.

Other findings will be described in detail in a forthcoming publication [8]. This includes the epithermal nature of antihydrogen formation when the positron density is high, a drift of antiprotons away from the axis of the trap and a mechanism for field-ionization of so-called giant dipole states of antihydrogen.

Acknowledgments

We wish to thank Professor Francis Robicheaux for making his computer programs available and for valuable discussions. We are grateful to the EPSRC (UK) for the support of our work under grants EP/D038707/1, EP/E048951/1, EP/G041938/1 and EP/D069785/1. We also gratefully acknowledge the use of the North West Grid and the UK National Grid Service in carrying out this work.

References

[1] Amoretti M *et al.* 2002 *Nature* **419** 456
[2] Gabrielse G *et al.* 2002 *Phys. Rev. Lett.* **89** 213401
[3] Andresen G *et al.* 2007 *Phys. Rev. Lett.* **98** 023402

[4] Gabrielse G *et al.* 2007 *Phys. Rev. Lett.* **98** 113002
[5] See articles in these proceedings.
[6] Robicheaux F 2008 *J. Phys. B: At. Mol. Opt. Phys.* **41** 192001
[7] Nersisyan H B, Walter M and Zwicknagel G 2000 *Phys. Rev. E* **61** 7022
[8] Jonsell S, van der Werf D P, Charlton M and Robicheaux F Accepted for publication by *J. Phys. B: At. Mol. Opt. Phys.*
[9] Dubin D H E and O'Neil T M 1999 *Rev. Mod. Phys.* **71** 87
[10] Mansbach P and Keck J 1969 *Phys. Rev.* **181** 275
[11] Robicheaux F and Hanson J D 2004 *Phys. Rev. A* **69** 010701(R)
[12] Glinsky M E and O'Neil T M 1991 *Phys. Fluids B* **3** 1279
[13] Robicheaux F 2004 *Phys. Rev. A* **70** 022510
[14] Fujiwara M C *et al.* 2008 *Phys. Rev. Lett.* **101** 053401

Journal of Physics: Conference Series **199** (2010) 012009 doi:10.1088/1742-6596/199/1/012009

AEGIS: an experiment to measure the gravitational interaction between matter and antimatter

Michael Doser (for the AEGIS collaboration [1])
CERN EP, 1211 Geneva 23, Switzerland
E-mail: michael.doser@cern.ch

Abstract. Considerable efforts have been made and are still being made to verify the validity of the principle of the equivalence of gravitational and inertial mass, one of the cornerstones of the classical theory of general relativity. Specific attempts at quantum-mechanical formulations of gravity allow for non-Newtonian contributions which might lead to a difference in the gravitational force on matter and antimatter. While it is widely expected that the gravitational interaction is independent of the composition of bodies, this has only been tested for matter systems, but never yet for antimatter systems. By combining techniques from different fields, and relying on recent developments in the production of Positronium and ongoing work to laser-excite Positronium to Rydberg states, such a test with neutral antimatter has become feasible. The primary goal of the AEGIS experiment being built at the Antiproton Decelerator at CERN is to carry out the first direct measurement of the Earth's gravitational acceleration on antihydrogen by means of a classical Moiré deflectometer.

1. Introduction

A hypothetical quantum theory of gravitation necessarily constitutes a departure from the Einsteinian view of gravity as a geometric phenomenon and could potentially constitute a violation of the weak equivalence principle. Attempts at formulating a quantum theory of gravity have mainly been hampered by the fact that such a theory is non-renormalizable, though a renormalization within the framework of an effective field theory may turn out to be feasible [2]. As in other quantum field theories, the interaction is mediated by exchange particles. The spins of these exchange bosons as well as the signs of the charges to which they couple determine whether a force is repulsive or attractive. Modern theories of gravity that attempt to unify gravity with the other forces of nature allow that, at least in principle, antimatter may fall differently from normal matter in the Earth's gravitational field. Specifically, as pointed out by Sherk [3], N=2,..,8 theories of supergravity (where N is the number of supersymmetries) lead to the possibility of (massive, and thus finite range) Kaluza-Klein graviscalar and gravivector components to the gravitational interaction which may lead to different couplings for matter and antimatter.

While it is possible to constrain deviations from standard gravity through e.g. effects of virtual antiparticles in ordinary matter, most of the quantitative limits [4, 5, 6] do not apply to more elaborate models involving vector and scalar gravitons [7]. If the hypothetical vector and scalar charges, as well as their masses (and thus the ranges of the interactions) are carefully chosen, such contributions can be strongly suppressed in ordinary matter. The most stringent limits on such additional components come from an analysis of the Eöt-Wash experiment

[8], an Eötvös-type torsion-balance experiment using test bodies of different compositions. Unlike ordinary matter [9], the behavior of antimatter particles in a gravitational field has never been tested experimentally. Two attempts, at Stanford [10] and CERN's Low-Energy Antiproton Ring [11] were thwarted by the overwhelming effect of stray electric and magnetic fields upon the electrically charged test particles. The recent production of copious amounts of cold antihydrogen $\bar{H}$ at CERN's Antiproton Decelerator (AD) [12, 13] has paved the way for high-precision gravity experiments with neutral antimatter. We have proposed the AEGIS experiment (Antimatter Experiment: Gravity, Interferometry, Spectroscopy), to be performed at CERN/AD, in order to address this important question.

The primary scientific goal of AEGIS is the direct measurement of the Earth's local gravitational acceleration g on $\bar{H}$. In a first phase of the experiment, a gravity measurement with 1% relative precision will be carried out by observing the vertical displacement (using a high-resolution position sensitive detector) of the shadow image produced by an $\bar{H}$ beam, formed by its passage though a Moiré deflectometer, the classical counterpart of a matter wave interferometer. This measurement will represent the first direct determination of the effect of gravity on antimatter.

The essential steps leading to the production of a pulsed cold beam of $\bar{H}$ and the measurement of g with AEGIS are the following (Fig. 1):

- Production of positrons (e^+) from a Surko-type source and accumulator;
- Capture and accumulation of $\bar{p}$ from the AD in a cylindrical Penning trap;
- Cooling of the $\bar{p}$ to sub-K temperatures
- Production of positronium (Ps) by bombardment of a cryogenic nanoporous material with an intense e^+ pulse;
- Excitation of the Ps to a Rydberg state with principal quantum number n $= 30 \sim 40$;
- Pulsed formation of $\bar{H}$ by resonant charge exchange between Rydberg Ps and cold $\bar{p}$;
- Pulsed formation of an $\bar{H}$ beam by Stark acceleration with inhomogeneous electric fields;
- Determination of g in a two-grating Moiré deflectometer coupled with a position-sensitive detector.

The feasibility of the first two points has been conclusively demonstrated by the ATHENA and ATRAP collaborations (see, in particular [14, 15]). In the following, we will discuss the remaining aspects of the proposed technique in more detail.

2. Method

2.1. Positronium production and excitation

In recent years, the potential for nanoporous insulator materials to be used as highly efficient Ps converters has been recognized, and the relevant formation mechanisms have been studied extensively [16]. When e^+ are implanted into such a material at kinetic energies ranging from several 100 eV to a few keV, they scatter off atoms and electrons in the bulk and are slowed to eV energies within a few ps. With efficiencies ranging from 10% to 50%, the slow e^+ capture either bound e^- or those liberated in prior collisions and form Ps. In the pores, Ps repeatedly bounces off the cavity walls and eventually approaches complete thermalization with the target material.

While some ortho-Ps are lost due to so-called pick-off annihilations of e^+ with the molecular e^- of the cavity walls, a sizable fraction diffuses out of the film at thermal energies. The overall ortho-Ps yield as well as the final velocity distribution depend upon the characteristics of the target material (in particular, its pore structure), the implantation depth, and the target temperature. For the AEGIS experiment, the precise degree of thermalization is critical, since at too low a Ps temperature, annihilation in the target material will dominate, while at too high

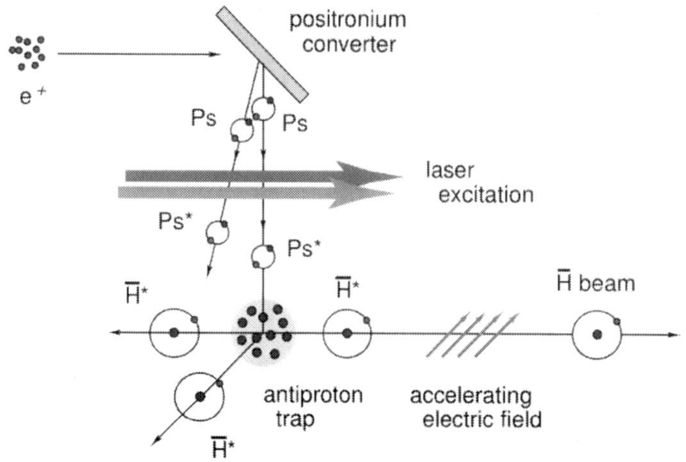

Figure 1. Proposed method for the production of a pulsed beam of cold $\bar{H}$ atoms.

a Ps temperature, the subsequent charge exchange cross section (to form $\bar{H}$) drops rapidly [17]. By carefully tailoring the topology of the target material's pores however, a degree of control appears possible [18].

Measurements using positron annihilation lifetime spectroscopy have shown that the ortho-Ps fraction (outside the sample) can exceed 20% in silicon-based polymer materials cooled to 50 K [19]. This is shown in Fig. 2 , where the ortho-Ps fraction is displayed as a function of implantation energy for different target materials and temperatures. In other experimental work, it was shown that the energy profile of Ps emitted from the surface of a silica film at room temperature followed a Maxwell-Boltzmann distribution and was compatible with the Ps being fully thermalized [20]. We are currently conducting experiments in order to determine the optimal converter material and e^+ energy in terms of ortho-Ps yield. Furthermore, we are investigating how well the emitted Ps is thermalized at very low target temperatures.

The photo-excitation of Ps to Rydberg states requires photon energies close to the binding energy of 6.8 eV. Laser systems at the corresponding wavelengths (180 nm) are not commercially available. We are therefore planning to perform a two-step excitation, from the ground state to the n = 3 state (λ = 205 nm), and then to the n = 35 Rydberg band ($\lambda \sim 1670nm$) [21]. Two pulsed-laser systems, both of which are pumped by a Q-switched Nd:YAG laser (1064 nm, 4 ns, 140 mJ), are currently under development. Both systems must provide sufficient power to excite the emitted Ps within a few ns and must be geometrically matched to the expanding cloud. Furthermore, the bandwidths of the lasers must be tailored to the transition linewidth broadened due to the Doppler effect as well as level splitting due to the motional Stark effect and the linear and quadratic Zeeman effect.

2.2. *Antihydrogen recombination and beam formation*

An $\bar{H}$ recombination scheme based on resonant charge exchange with Ps was first proposed almost twenty years ago [22]. The reaction proceeds according to the equation

$$Ps^* + \bar{p} \rightarrow \bar{H}^* + e^- \qquad (1)$$

where the star denotes a highly excited Rydberg state. This reaction owes its appeal to the fact that the cross-section scales approximately with the fourth power of the principal quantum

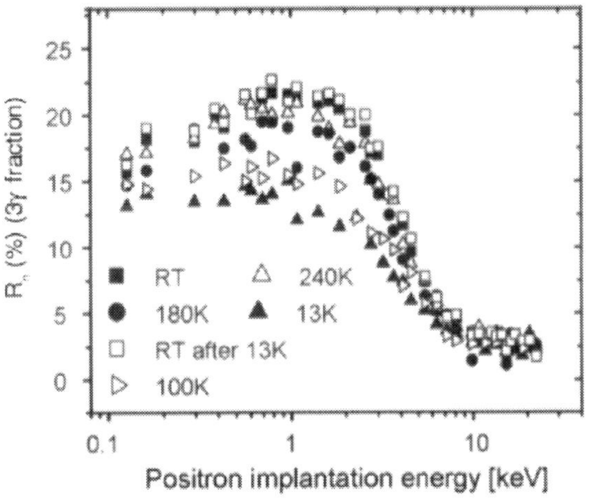

Figure 2. Ortho-Ps fraction as a function of implantation energy into a closed porosity (top figure) and open porosity (bottom figure) target for different target temperatures between 13K and 200K.

number. In addition, it creates $\bar{H}$ in a narrow and well-defined band of final states. Most importantly, $\bar{H}$ formed with $\bar{p}$ at rest is created with a velocity distribution dominated by the $\bar{p}$ temperature, hence the surrounding (cryogenic) environment [23]. This is in stark contrast to the rather high $\bar{H}$ temperature observed when using the nested-well technique pioneered by ATRAP and ATHENA [24, 25]. Our proposed technique is conceptually similar to a charge exchange technique based on Rydberg cesium [26] which has been successfully demonstrated by ATRAP [27], but offers greater control of the final state distribution of $\bar{H}$ and, more importantly, allows pulsed production of $\bar{H}$.

The principle is illustrated in Fig. 1. The Ps emitted from the porous insulator material are excited to Rydberg states. They then traverse a Penning trap region in which $\sim 10^5 \bar{p}$ have been accumulated, stored and cooled to O(100 mK). The low temperature requirement on the antiprotons comes from the requirement that the antihydrogen atoms that will be formed will have a velocity that is low compared to the velocity of several 100 m/s that they will achieve after acceleration. To reach such a low temperature, the Penning trap is coupled to a 50 mK dilution refrigerator, and the antiprotons are coupled to the low temperature environment by embedding them in an electron plasma. The later will cool through synchrotron radiation, as well as through a tuned circuit; furthermore, evaporative cooling of the pre-cooled antiprotons is being envisaged. The charge exchange cross-section is very large ($\sim 10^7 \mathring{A}^2$ for n = 35) and reaches a maximum when the e^+ and $\bar{p}$ relative velocities are matched. Taking into account the corresponding kinetic energy, as well as a smaller contribution due to converted internal energy, $\bar{H}$ is created at velocities of $25 \sim 80\ ms^{-1}$.

While neutral atoms are not sensitive (to first order) to constant electric fields, they do experience a force when their electric dipole moment is exposed to an electric-field gradient. Since the dipole moment scales approximately with the square of the principal quantum number, Rydberg atoms are especially amenable to being manipulated in this way. Such so-called Stark acceleration (and deceleration) has been successfully demonstrated, among others, by one of the AEGIS groups with (ordinary) hydrogen after excitation to the n = 22, 23, 24 states [28, 29]. In these experiments, accelerations of $2 \times 10^8 ms^{-2}$ were achieved using two pairs of electrodes

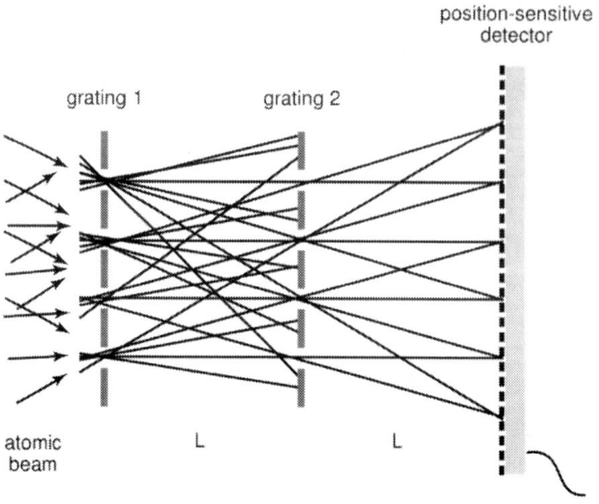

Figure 3. Schematic of the principle of the Moiré deflectometer consisting of two gratings coupled to a position-sensitive detector.

at right angles to each other. A hydrogen beam traveling at 700 ms^{-1} was stopped within 5 μs over a distance of only 1.8 mm. We intend to use a similar field configuration, generated by axially split electrodes within the cylindrical geometry of a Penning trap, to accelerate the formed $\bar{H}$ atoms to about 400 ms^{-1} in the direction of the deflectometer apparatus.

2.3. *Gravity measurement*

In matter wave interferometers of the Mach-Zehnder type [30, 31], three identical gratings are placed at equal distances L from each other. The first two gratings produce an interference pattern at the location of the third. That pattern has the same period d as the gratings, and its position perpendicular to the diffracted particle beam can be determined precisely by displacing the third grating and recording the overall transmission with a particle detector. Under the influence of gravity, the interference pattern is vertically displaced (it falls) by a distance

$$\delta x = -gT^2 \tag{2}$$

where g is the local gravitational acceleration and T is the time of flight L/v between each pair of gratings of a particle beam traveling at velocity v.

Contrary to such true interferometers, which place a very stringent limit on the acceptable beam divergence (and thus antihydrogen temperature at production), the so-called Moiré deflectometer, in which diffraction on the gratings is replaced by a (classical) shadow pattern of those particles that converge onto the third grating, works in the classical regime. Interestingly, the gross characteristics of the interferometer are retained [32], in particular, the vertical displacement of the interference pattern according to Eq. (2). A three-grating Moiré deflectometer has been used to measure the local gravitational acceleration to a relative precision of 2×10^{-4} with a beam of argon atoms traveling at an average velocity of 750 ms^{-1} [32]. In departing from the three-grating deflectometer, we intend to replace the third grating by a position-sensitive silicon strip detector (see Fig. 3). Antihydrogen atoms impacting on the detector plane annihilate, and the impact point can be reconstructed by means of the energy deposited locally by the annihilation products.

The value of g is extracted from two primary observables (time of flight T and vertical displacement of the fringe pattern δx). The periodic nature of the arrangement means that for a given value of T, the impact point will have dropped by a well-defined amount $\delta x(T)$, modulo the grating period. By varying the accelerating field gradient (and thus varying the velocity of the antihydrogen atoms), or simply through the spread in velocity of the ensemble of antihydrogen atoms, the quadratic dependence of δx on T is probed.

Our simulations have shown that in order to perform a measurement of g to 1% relative precision, about 10^5 H atoms at a temperature of 100 mK will be required, equivalent to several months of data taking at the AD. In these simulations, a grating period of 80 μm was used, and a finite detector resolution of 10 μm was taken into account.

3. Summary and outlook

Construction has started on the AEGIS experiment, whose design is based upon the broad experience gained with the ATHENA and ATRAP experiments at the AD, a series of ongoing related experiments, tests and developments, as well as extensive simulations of critical processes (charge exchange production of $\bar{H}$, Stark acceleration and propagation through the Moiré deflectometer, resolution of the position-sensitive detector located at the end of the deflectometer). The proposed gravity measurement merges in a single experimental apparatus technologies already demonstrated or based on reasonable additional development.

For the initial phase of the experiment, obtaining samples of anti-atoms at 100 mK is an essential requirement. Gravity measurements with even higher precision, as well as competitive CPT tests through spectroscopy, are desirable, but will necessitate the development of novel techniques to attain even colder $\bar{H}$ ensembles. The experiment has been designed with flexibility of the apparatus in mind, in order to allow a number of techniques, which may lead to such physics topics, to be implemented. One natural extension of the modular design is to incorporate, in a future stage, a magnetic decelerator and trap for $\bar{H}$, which will be spatially separated from the region where the anti-atoms are produced, similar to the devices currently being used to trap and study H atoms [33, 34]. The experience gained in the first phase of AEGIS with the formation of an $\bar{H}$ beam will be used to optimize the design of such a trapping system.

[1] http://aegis.web.cern.ch/aegis
[2] J.F. Donoghue, Phys. Rev. D 50 (1994) 3874.
[3] J. Sherk, Phys. Lett. B 88 (1979) 265
[4] L.I. Schiff, Phys. Rev. Lett. 1 (1958) 254.
[5] L.I. Schiff, Proc. Natl. Acad. Sci. 45 (1959) 69. (http://www.pnas.org/cgi/reprint/45/1/69)
[6] R.J. Hughes, M.H. Holzscheiter, Phys. Rev. Lett. 66 (1991) 854.
[7] M.M. Nieto, T. Goldman, Phys. Rep. 205 (1991) 221.
[8] D. Alves, M. Jankowiak, P. Saraswat, arXiv:0907.4110v1
[9] B. Heckel et al., Adv. Space Res. 25 (2000) 1225
[10] W.M. Fairbank et al., in: B. Bertotti (Ed.), Proc. Int. School Phys. Enrico Fermi, Academic Press, New York, 1974, p. 310.
[11] M.H. Holzscheiter et al., Nucl. Phys. A 558 (1993) 709c.
[12] M. Amoretti et al., ATHENA Collaboration, Nature 419 (2002) 456.
[13] G. Gabrielse et al., ATRAP Collaboration, Phys. Rev. Lett. 89 (2002) 213401.
[14] L.V. Jorgensen et al., ATHENA Collaboration, Phys. Rev. Lett. 95 (2005) 025002.
[15] G. Gabrielse et al., ATRAP Collaboration, Phys. Lett. B 548 (2002) 140.
[16] D.W. Gidley, H.-G. Peng, R.S. Vallery, Annu. Rev. Mater. Res. 36 (2006) 49
[17] C. Canali, publication in preparation.
[18] S. Mariazzi, A. Salemi, R. Brusa, Phys. Rev. B 78 (2008) 085428.
[19] S. Mariazzi, A. Salemi, R. Brusa, Appl. Sruf. Sci. 255 (2008) 191.
[20] R.S. Vallery, P.W. Zitzewitz, D.W. Gidley, Phys. Rev. Lett. 90 (2003) 203402
[21] F. Castelli, I. Boscoli, S. Cialdi, D. Comparat, Phys. Rev. A 78 (2008) 052512.
[22] M. Charlton, Phys. Lett. A 143 (1990) 143.
[23] B.I. Deutch et al., Hyperfine Interact. 76 (1993) 153.

Journal of Physics: Conference Series **199** (2010) 012009 doi:10.1088/1742-6596/199/1/012009

[24] G. Gabrielse et al., ATRAP Collaboration, Phys. Rev. Lett. 93 (2004) 073401
[25] N. Madsen et al., ATHENA Collaboration, Phys. Rev. Lett. 94 (2005) 033403
[26] E.A. Hessels, D.M. Homan, M.J. Cavagnero, Phys. Rev. A 57 (1998) 1668
[27] C.H. Storry et al., ATRAP Collaboration, Phys. Rev. Lett. 93 (2004) 263401
[28] E. Vliegen, F. Merkt, J. Phys. B 39 (2006) L241.
[29] E. Vliegen et al., Phys. Rev. A 76 (2007) 023405.
[30] L. Zehnder, Z. Instrumentenkunde 11 (1891) 275.
[31] L. Mach, Z. Instrumentenkunde 12 (1892) 89.
[32] M.K. Oberthaler et al., Phys. Rev. A 54 (1996) 3165.
[33] S. Hogan and F. Merkt, Phys. Rev.Lett. 100 (2008) 043001
[34] S. Hogan, A. Wiederkehr, H. Schmutz and F. Merkt, Phys. Rev. Lett. 101 (2008) 143001

Journal of Physics: Conference Series **199** (2010) 012010 doi:10.1088/1742-6596/199/1/012010

Leading-order relativistic corrections to the Zeeman splitting of hyperfine structure levels in two-fermion bound-state systems

Andrei G. Terekidi[a], Jurij W. Darewych[b], Marko Horbatsch[c]

Department of Physics and Astronomy, York University, Toronto, Ontario, M3J 1P3, Canada
[a]terekidi@yorku.ca, [b]darewych@yorku.ca,[c]marko@yorku.ca

Abstract

The relativistic calculations of the Zeeman splitting of hyperfine levels in two-fermion systems (hydrogen, anti-hydrogen , muonium, muonic hydrogen) are presented. Our approach is based on the variational equation for bound states derived from quantum electrodynamics [1]. Relativistic corrections to the g-factor are obtained up to $O\left((\alpha)^2\right)$. Calculations are provided for all quantum states and for arbitrary fermionic mass ratio. The results will be useful for comparison with high-precision measurements.

1. Introduction

In a recent paper [2] we have presented a self-consistent variational method for calculating the non-relativistic Landé g factor of the two-fermion bound-state system. In the lowest-order approximation the linearly dependent part of the energy splitting for a two-fermion system placed in a weak static magnetic field $\mathbf{B}$ can be written as [2-4]

$$\Delta E^{ext}_{J,m_J,S,\ell,s_1,s_2} = (\mu_{B1}g_1 + \mu_{B2}g_2)\, Bm_J, \tag{1}$$

where J, m_J, S, ℓ, s_1, s_2 are quantum numbers, which characterize the system: s_1 and s_2 are the spins of the first and second particle respectively, $S = s_1 + s_2,\ s_1 + s_2 - 1, ..., \ |s_1 - s_2|$ is the total spin of the particles, ℓ and J represent the orbital and total angular momentum quantum numbers, where $J = \ell + S,\ \ell + S - 1, ...,\ |\ell - S|$. The projection of the total angular momentum on the $\mathbf{B}$ field direction is labeled by $m_J = -J,\ -J+1, ...,\ J-1,\ J$. The "Bohr magnetons" for the two particles are defined as $\mu_{B1} = Q_1\hbar/2m_1c$, and $\mu_{B2} = -Q_2\hbar/2m_2c$, where $Q_1 = Z_1e$ and $Q_2 = Z_2e$ represent the magnitude of the charges ($e > 0$). In our notation, m_1 and m_2 correspond to the masses of the light and heavy particle respectively. The description of the interacting system by the set of quantum numbers J, m_J, S, ℓ, s_1, s_2 corresponds to the LS coupling representation. This representation is used in contrast to the customary j_1-j_2 coupling scheme (for the case $m_2 >> m_1$), where the states are taken to be the eigenstates of the operators $\widehat{\mathbf{j}}_1^2 = \left(\widehat{\mathbf{L}} + \widehat{\mathbf{s}}_1\right)^2$, $\widehat{\mathbf{j}}_2^2 = \widehat{\mathbf{s}}_2^2$. As discussed in [1,2,5,6], for the general case of arbitrary mass ratio, the j_1 value is not a good quantum number. Even in the LS representation, for spin-mixed states the orbital angular momentum L, and total spin S of the system are not conserved. In this case we designate the states by an additional quantum number s_q, which takes on the values of 0 or 1 for quasisinglet (sg_q) and quasitriplet (tr_q) states respectively.

In our calculations we assume that the energy-level splitting (1) is smaller than the hyperfine structure (HFS) splitting, $\Delta E^{ext} << \Delta E^{HFS}$, i.e., we treat the interaction with an external magnetic field as a perturbation.

The g-factors in (1) can be written more generally as $g_i = g_i^{NR} + \triangle g_i^{REL}$, where g_i^{NR} are the nonrelativistic Landé factors obtained in [2], and $\triangle g_i^{REL}$ are the relativistic correction. In the next section we calculate $\triangle g_i^{REL}$ to order $O\left((Z_1 Z_2 \alpha)^2\right)$ for all quantum states and arbitrary masses of particles. In most expressions we use natural units, i.e., $\hbar = c = 1$, $\alpha = e^2/4\pi$. The coupling constant is defined as $\alpha' = Z_1 Z_2 \alpha$.

2. Relativistic corrections to g-factors of two-fermion systems

The relativistic wave equations for two-fermion systems in the absence of external fields were derived in [1] and [6] on the basis of a modified (reformulated) QED Lagrangian [8]-[10]. In this approach a simple Fock-space trial state taken in the rest frame is $|\psi_{trial}\rangle = \sum_{s_1 s_2} \int d^3\mathbf{p} F_{s_1 s_2}(\mathbf{p}) b^\dagger_{-\mathbf{p}s_1} D^\dagger_{\mathbf{p}s_2} |0\rangle$. This form is sufficient to obtain the HFS levels correct to fourth order in the coupling constant α'. Here $b^\dagger_{\mathbf{q}_1 s_1}$ and $D^\dagger_{\mathbf{q}_2 s_2}$ are creation operators for a free fermion of mass m_1 and an (anti) fermion of mass m_2 respectively, and $|0\rangle$ is the trial vacuum state such that $b_{\mathbf{q}_1 s_1}|0\rangle = D_{\mathbf{q}_2 s_2}|0\rangle = 0$. The $F_{s_1 s_2}$ are four-component adjustable functions. The trial state is taken to be an eigenstate of total linear momentum $\widehat{\mathbf{P}} = 0$, total angular momentum squared $\widehat{\mathbf{J}}^2$, its projection $\widehat{J}_3$, parity $\widehat{\mathcal{P}}$, and the Hamiltonian $\widehat{H}$, which corresponds to the hyperfine interaction [6]. The adjustable functions $F_{s_1 s_2}(\mathbf{p})$ can be specified for two categories:

(i) The spin-mixed (quasi-singlet and quasi-triplet) states

In this case we have $\ell = J$, and the general solution under the condition of well-defined eigenvalues of $\widehat{\mathbf{P}}$, $\widehat{\mathbf{J}}^2$, $\widehat{J}_3$, and $\widehat{\mathcal{P}}$ can be expressed as [1], [6]

$$F_{s_1 s_2}(\mathbf{p}) = C^{(S_1) m_{s_1 s_2}}_{J m_J} f_J^{(S_1)}(p) Y_J^{m_{s_1 s_2}}(\widehat{\mathbf{p}}) + C^{(S_2) m_{s_1 s_2}}_{J m_J} f_J^{(S_2)}(p) Y_J^{m_{s_1 s_2}}(\widehat{\mathbf{p}}), \tag{2}$$

where $C^{(S_{1,2}) m_{s_1 s_2}}_{J m_J} = \langle \ell m_\ell S m_S \mid J m_J \rangle$ are the Clebsch-Gordan (CG) coefficients with total spin S, where $S = 0$ (with index S_1) for the singlet states and $S = 1$ (with index S_2) for the triplet states respectively. Note that $m_{11} = 1$ if $m_S = 1$, $m_{12} = m_{21} = 0$ if $m_S = 0$, and $m_{22} = -1$ if $m_S = -1$. Here $f_J^{(S_1)}(p)$ and $f_J^{(S_2)}(p)$ are radial functions to be determined. They represent the contributions of spin-singlet and spin-triplet states (the total spin $S = 0, 1$ is not conserved in general).

(ii) The ℓ-mixed triplet states

These states occur for $\ell_{1,2} = J \mp 1$. Their radial decomposition can be written as

$$F_{s_1 s_2}(\mathbf{p}) = C^{(\ell_1) m_{s_1 s_2}}_{J m_J} f_{\ell_1}(p) Y_{\ell_1}^{m_{s_1 s_2}}(\widehat{\mathbf{p}}) + C^{(\ell_2) m_{s_1 s_2}}_{J m_J} f_{\ell_2}(p) Y_{\ell_2}^{m_{s_1 s_2}}(\widehat{\mathbf{p}}). \tag{3}$$

Again, the $C^{(\ell_{1,2}) m_{s_1 s_2}}_{J m_J} = \left\langle \ell_{1,2} m_{\ell_{1,2}} S m_S \mid J m_J \right\rangle$ are CG coefficients. For these states the system is characterized by J, m_J, and $\mathcal{P} = (-1)^J$. The orbital angular momentum $\ell = \ell_{1,2}$ is not a good quantum number. Mixing of this type occurs only for states with principal quantum number $n \geq 3$.

From the variational principle $\delta\left\langle \psi_{trial}\left|\widehat{H}-E\right|\psi_{trial}\right\rangle=0$, where $\widehat{H}$ is the QED Hamiltonian is invoked to obtain a momentum-space wave equation [1], we get a system of coupled radial equations expressed in matrix form as

$$\left(\omega_p+\Omega_p-E\right)\mathbb{F}\left(p\right)=\frac{m_1m_2}{\left(2\pi\right)^3}\int\frac{q^2dq}{\sqrt{\omega_p\omega_q\Omega_p\Omega_q}}\mathbb{K}\left(p,q\right)\mathbb{F}\left(q\right), \tag{4}$$

where $\omega_p^2=\mathbf{p}^2+m_1^2$ and $\Omega_p^2=\mathbf{p}^2+m_2^2$, and $q=|\mathbf{q}|$. Here $\mathbb{F}\left(p\right)$ is the two-component matrix of radial functions $\mathbb{F}^{\dagger}\left(p\right)=\left[\begin{array}{cc} f_J^{(S_1)}(p) & f_J^{(S_2)}(p)\end{array}\right]^*$, $\left[\begin{array}{cc} f_{\ell_1}(p) & f_{\ell_2}(p)\end{array}\right]^*$ for spin-mixed and ℓ-mixed states respectively. The kernel of this equation is the 2×2 matrix $\left[\mathbb{K}\right]_{ij}=\mathcal{K}_{ij}$, which has the following form

$$\mathcal{K}_{ij}=-i\sum_{s_1s_2\sigma_1\sigma_2}C_{Jm_Jij}^{s_1s_2\sigma_1\sigma_2}\int d\widehat{\mathbf{q}}d\widehat{\mathbf{p}}\;Y_{\ell_i}^{m_{s_1s_2}*}\left(\widehat{\mathbf{p}}\right)Y_{\ell_j}^{m_{\sigma_1\sigma_2}}\left(\widehat{\mathbf{q}}\right)\widetilde{\mathcal{M}}_{s_1s_2\sigma_1\sigma_2}\left(\mathbf{p}_1,\mathbf{p}_2,\mathbf{q}_1,\mathbf{q}_2\right). \tag{5}$$

Here the $C_{Jm_Jij}^{s_1s_2\sigma_1\sigma_2}$ are related to the CG coefficients by: $C_{Jm_Jij}^{s_1s_2\sigma_1\sigma_2}=C_{Jm_J}^{(S_i)m_{s_1s_2}}C_{Jm_J}^{(S_j)m_{s_1s_2}}$ and $C_{Jm_Jij}^{s_1s_2\sigma_1\sigma_2}=C_{Jm_J}^{(\ell_i)m_{s_1s_2}}C_{Jm_J}^{(\ell_j)m_{\sigma_1\sigma_2}}$ for the spin- and ℓ-mixed states respectively. For the spin-mixed states we should take $\ell_i\equiv\ell_j\equiv\ell$. The inter-particle interaction is represented by the generalized invariant matrix $\widetilde{\mathcal{M}}=\sum_{n=1}\mathcal{M}^{(n)}$, obtained as part of the derivation. It includes reducible and irreducible effects in all orders of the coupling constant α', and the sum contains all relevant Feynman diagrams. A discussion of the derivation and structure of the $\widetilde{\mathcal{M}}$-matrix to one-loop level is provided in [1]. This equation allows one to obtain, in principle, all relativistic and QED corrections to the g-factor.

The lowest-order QED corrections appear within the term $\mathcal{M}^{(2)}$ and can be formally included in the intrinsic factor g_{s_i}, however we shall not do so in this work. In this paper we restrict our consideration to the first term $\mathcal{M}^{(1)}$ of the expansion, i.e., only tree level diagrams are included. The term $\mathcal{M}^{(1)}$ contains only relativistic corrections and it can be broken into two parts, namely $\mathcal{M}^{(1)}=\mathcal{M}^{ope}+\mathcal{M}^{ext}$, where $\mathcal{M}^{ope}$ is the usual invariant matrix element, corresponding to the one-photon exchange Feynman diagram [1]. The element $\mathcal{M}^{ext}$ represents the interaction with a given external classical field A_μ^{ext},

$$\mathcal{M}^{ext}=\left(2\pi\right)^{3/2}\left(\begin{array}{c}\frac{\sqrt{\Omega_{\mathbf{p}_2}\Omega_{q_2}}Q_1}{m_2}A_\mu^{ext}(\mathbf{p}_1-\mathbf{q}_1)\overline{u}\left(\mathbf{p}_1,s_1\right)\gamma^\mu u\left(\mathbf{q}_1,\sigma_1\right)\delta_{s_2\sigma_2}\\ +\frac{\sqrt{\omega_{\mathbf{p}_1}\omega_{\mathbf{q}_1}}Q_2}{m_1}A_\mu^{ext}(\mathbf{q}_2-\mathbf{p}_2)\overline{V}\left(\mathbf{p}_2,\sigma_2\right)\gamma^\mu V\left(\mathbf{q}_2,s_2\right)\delta_{s_1\sigma_1}\end{array}\right). \tag{6}$$

The non-relativistic limit of this matrix element was considered in [2]. Using the relativistic expansion of the expression $\overline{u}\left(\mathbf{p}_1,s_1\right)\gamma u\left(\mathbf{q}_1,\sigma_1\right)$ up to order $1/c^3$ we obtain the lowest-order relativistic correction to $\mathcal{M}^{ext}$. The solution of equation (4) including only the first two terms with the $\mathcal{M}^{ope}$ and $\mathcal{M}^{(1)ext}$ matrices was discussed in [2]. This solution describes the Zeeman splitting of the HFS energy levels in the non-relativistic limit. In order to obtain the Landé factors to order $O\left(\alpha'^2\right)$ we solve the radial equation (4) with the additional term $\mathcal{M}^{(2)ext}$ in the kernel (5), which is evaluated perturbatively using hydrogen-like radial functions in momentum space [4].

The calculations are straightforward, and yield the relativistic corrections to the g-factor for both particles of the system. We obtain for the triplet states

$$\triangle g_i^{REL} = -\frac{\mu_j^2}{2}\left(\mu_j + \frac{\mu_i}{l'} - \frac{1}{2J+1}\right)\left(\frac{\alpha'}{n}\right)^2, \tag{7}$$

where $l' = J$, when $l = J-1$, and $l' = J+1$, when $l = J+1$. For the spin-mixed states we have ($l = J$)

$$\triangle g_i^{REL} = -\mu_j^2\left(\mu_j + \mu_i\frac{1 \mp \xi}{4J(J+1)}\right)\left(\frac{\alpha'}{n}\right)^2. \tag{8}$$

The mass factors μ_i are defined as $\mu_i = m_i/(m_1+m_2)$, where $i = 1,2$ is the index designating the particle. The index j is defined as $j = 1$ when $i = 2$, and $j = 2$ when $i = 1$. The parameter ξ is defined as $\xi = \left(4(\mu_1-\mu_2)^2 J(J+J)+1\right)^{-1/2}$ The upper and lower sign in (8) corresponds to sg_q and tr_q states respectively. Our result (7)-(8) is applicable to a system of two equal masses (positronium). However, up to the order we consider here, due to a symmetry of the system like positronium ($\mu_{Be^-}g_{e^-} = -\mu_{Be^+}g_{e^+}$), the total effect of the energy splitting (1) is zero.

There are several works based on the Breit equation (e.g. [11]-[15]), where the relativistic corrections to the g-factor of two-fermion bound states were considered. Some of the calculations were obtained only for small mass ratio [11]; in others the orbital motion of the heavy particle was ignored [12]. These calculations did not take into account the mixed nature of the quantum states. We emphasize, that only mixed states diagonalize the Hamiltonian of HFS [2], [6]. Our result (7)-(8) is new and overcomes the above-mentioned shortcomings. For the relatively simple case of the ground state $1S_{1/2}$ ($J = 1$, $\mathcal{P} = -1$) our results agree with the previous result of Hegstrom [13] and Grotch [14], namely $\triangle g_i^{REL} = -\mu_j^2\alpha'^2/3$. Note that their definition of g_e based upon $\Delta E_e = \mu_{B1}g_e B m_s$, Eq. (12) of [13], differs by a factor of 2 from our definition (1).

We provide our result (7), (8) in numerical form for the lighter particle in atomic hydrogen, muonium, and muonic hydrogen, for which $\alpha' = \alpha$. We consider only states with principal quantum number $n = 1,2$. We used the following values for the mass ratios: $m_e/m_p = 5.4461702\times10^{-4}$ and $m_p/m_\mu = 8.8802433$. The fine-structure constant is taken as $\alpha = 7.2973523\times10^{-3}$.

Table. The relativistic correction $\triangle g_1^{REL}$ to the Landé factor for $n = 1,2$ states in hydrogen, muonium and muonic hydrogen.

	$1S_{1/2(J=1)}$	$2S_{1/2(J=1)}$	$2P_{1/2(J=1)}$	$2P_{3/2(J=1)}$	$2P_{3/2(J=2)}$
pe^-	−0.0000177	−0.0000044	−0.0000133	−0.0000133	−0.0000053
μ^+e^-	−0.0000176	−0.0000044	−0.0000131	−0.0000131	−0.0000053
$p^+\mu^-$	−0.0000143	−0.0000036	−0.0000099	−0.0000098	−0.0000040

For the systems considered in the Table the relativistic corrections for the heavier particle $\triangle g_2^{REL}$ are negligible in comparison with $\triangle g_1^{REL}$ due to the small mass factor μ_1. For a realistic comparison of the g-factor with experiment one would also need to calculate the

QED corrections up to second order in α, that is the higher order matrix element $\mathcal{M}^{(2)}$ would have to be included.

In our calculations we assumed that the trial state is an eigenstate of the total linear momentum operator $\widehat{P}$. However this is only an approximation, because one can show that the operator $\widehat{P}$ does not commute with the HFS Hamiltonian. To fix this problem we need to modify the trial state, or use an appropriate unitary transformation for the HFS Hamiltonian. The latter approach was discussed in the literature [11], [12], [15]. An analysis shows, that in our case the unitary transformation leads to the appearance of additional terms in the invariant $\widetilde{\mathcal{M}}$-matrix. This is a technically difficult problem which we postpone for the future.

We note that, in contrast to the Breit approach, the anomalous magnetic moment is not introduced from another calculation. As discussed below equation (5) all QED effects are contained in the $\widetilde{\mathcal{M}}$-matrix. The anomalous magnetic moment will appear naturally in our calculations if we include the next term $\mathcal{M}^{(2)}$ of the expansion of $\widetilde{\mathcal{M}}$- matrix. The present work will be of practical importance when these measurements will be extended to all $n = 2$ (or higher-n) levels.

AGT and MH acknowledge the financial support of the Natural Sciences and Engineering Research Council of Canada for this work.

References

1. A. G. Terekidi, J. W. Darewych, M. Horbatsch, Can. J. Phys. **85**, 813-836.(2007).
2. A. G. Terekidi, J. W. Darewych, M. Horbatsch, Phys. Rev. A **75**, 043401 (2007).
3. V. W. Hughes and G. zu Putlitz, in *Quantum Electrodynamics*, edited by T. Kinoshita (World Scientific, Singapore, 1990), p. 822.
4. H. A. Bethe and E. E. Salpeter, *Quantum Mechanics of One- and Two-Electron Atoms* (Springer, 1957).
5. J. H. Connel, Phys. Rev. D **43**, 1393 (1991), H. Hersbach, Phys. Rev. A **46**, 3657 (1992).
6. A. G. Terekidi, J. W. Darewych, Journal of Mathematical Physics **45**, 1474 (2004), **46**, 032302 (2005).
7. J. Schwinger, Phys. Rev. **73**, 416 (1948).
8. W. T. Grandy, *Relativistic Quantum Mechanics of Leptons and Fields* (Kluwer, Berlin, 1991).
9. J. W. Darewych, Annales Fond. L. de Broglie (Paris) **23**, 15 (1998).
10. J. W. Darewych, in *Causality and Locality in Modern Physics*, G Hunter et al. (eds.), p. 333, (Kluwer, 1998).
11. H. Grotch, R. Kashuba, Phys. Rev. A, **7**, 78-84 (1973)
12. J. M. Anthony, K. J. Sebastian, Phys. Rev. A, **49**, 192-206 (1994)
13. R. A. Hegstrom, Phys. Rev. **184**, 1, p. 17-22 (1969).
14. H. Grotch, Phys. Rev. Let. **24**, 2, p.39-42 (1970).
15. H Grotch, R. A. Hegstrom, Phys. Rev. A, **4**, 59-69 (1971).

Excited, bound and resonant positron-atom systems

M W J Bromley[1] and J Mitroy[2]

[1] Department of Physics and Computational Science Research Center, San Diego State University, San Diego CA 92182, USA

[2] ARC Centre for Antimatter-Matter Studies and Faculty of Education, Health and Science, Charles Darwin University, Darwin NT 0909, Australia

E-mail: mbromley@physics.sdsu.edu

Abstract. Calculations have demonstrated that eleven neutral atoms can bind positrons, while many more can bind positronium. This is a short review of recent progress made in understanding some of the underlying mechanisms. The emphasis here being on configuration interaction calculations with excited state configurations. These have demonstrated the existence of a ^{2}P^o excited state of e^+Ca, which consists predominantly of a positronium cluster orbiting the Ca$^+$ ion in the $L = 1$ partial wave. Preliminary results are presented of excited state positron binding to a model alkali atom, where the excited ^{1}P^o states are stable over a limited region. Implications for the unnatural parity, ^{2,4}S^o, states of PsH, LiPs, NaPs and KPs are also discussed. The e^+Mg, e^+Cu, e^+Zn and e^+Cd systems show a lack of a ^{2}P^o excited state, each instead possessing a low-energy p-wave shape resonance of varying strength.

1. Introduction

Whilst a 'positronic atom' remains to be observed experimentally, calculations over the past decade, or so, have conclusively demonstrated that a wide-range of atomic systems can bind to positrons. This article highlights the progress made in our understanding of some of the underlying mechanisms involved since a review was published by the present authors in 2002 [1]. The emphasis here is on the insights gained from a series of large-scale configuration interaction (CI) and stochastic variational method (SVM) calculations that have been applied to various bound state systems and their scattering brethren with excited state configurations.

Preliminary results are presented of excited state positron binding to a model alkali atom. The ground ^{1}P^o states are shown to be be stable over a limited range of ionisation potentials, while the excited ^{1}S^e and the ground ^{1}P^e states appear to be unstable. These calculations were motivated, in part, by the Surko group measurements of positron binding energies in molecules which show extraordinary agreement with the atomic trend for the ground state ^{1}S^e model alkali atom [2]. The precise measurements have enabled the development of an empirical model that predicts ground state energies [3], and have guided a simple computational model that predicts the emergence of the second bound state with a node along the molecular axis at around the same energies as the experiment [4, 2].

2. Positron-atom binding mechanisms

The heuristic model for positron binding to atoms is simply based on the ionisation potential of the atom [5]. This dictates to what extent the positron can draw the electron outwards, thereby polarising the atom. A positron-atom wavefunction can be schematically represented as [6, 7, 8]

$$\Psi = c_0\,\Phi(\mathrm{Atom})\phi_{e^+}(\mathbf{r}_p) + c_1\,\Omega(\mathrm{Atom}^+)\omega_{\mathrm{Ps}}(\mathbf{R}_{\mathrm{Ps}}) + c_2\,\Xi(\mathrm{Atom}^{2+})\xi_{\mathrm{Ps}^-}(\mathbf{R}_{\mathrm{Ps}^-}) + \ldots \quad (1)$$

where the three components shown here represent a positron orbiting a polarised atom at a distance $\mathbf{r}_p$, a polarised positronium (Ps) cluster orbiting the residual atomic ion at $\mathbf{R}_{\mathrm{Ps}}$, and even a polarised positronium ion (Ps$^-$) cluster orbiting the atomic ion at $\mathbf{R}_{\mathrm{Ps}^-}$.

For the systems so far known to bind, being the group I/IB and group II/IIB atoms, only one or two electrons are active. The current best estimates of positron-atom binding energies are summarised in Figure 1. Shown in the figure is the result of a ^{1}S^e model positronic alkali atom, where an additional local potential acts on the valence electron only [5]. This provides a tuning-knob that changes the ionisation potential of the atom. This is discussed further shortly.

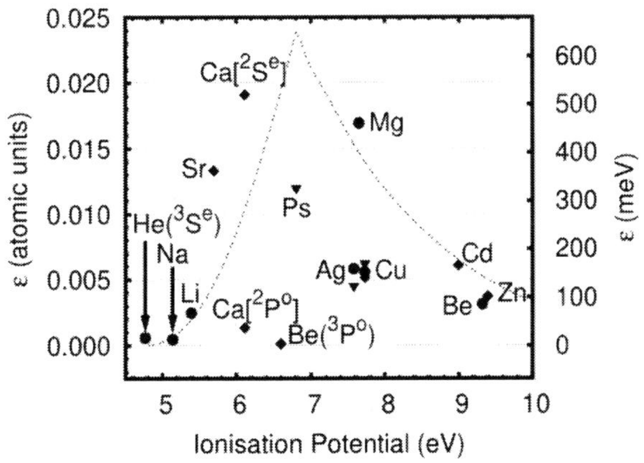

Figure 1. Theoretical e^+-atom binding energies versus the parent atomic ionisation energy [1]. The ● were computed with the SVM, the ♦ are CI$_{\mathrm{pol}}$-based, the ▲ are other calculations. The total $^{2S+1}L^\pi$ symmetries are implicit: the one-e^- atoms are ^{1}S^e, the two-e^- atoms are of ^{2}S^e. The (...) indicate the excited parent state symmetry, the [...] are the overall symmetry. The line is a model ^{1}S^e e^+-alkali atom system [5].

Each positronic atom wavefunction generally consists of a significant Ps-fraction when the ionisation potential lies to the left of the Ps atom energy $E_{\mathrm{Ps}} = 6.8$ eV. The computational representation of either a localised Ps cluster (or a significant component of Ps$^-$) in a wavefunction is computationally demanding [1]. In the extreme case of CI calculations of the ^{1}S^e e^+Li system, which has only one active electron, orbitals with angular momentum of up to $\ell = 30$ were required to demonstrate binding [9].

3. Excited state positron-atom binding mechanisms

Despite the computational challenges, the existence of a ^{2}P^o excited state of e^+Ca has been demonstrated. The structure consists predominantly of a Ps cluster orbiting the residual Ca$^+$ ion in the $L = 1$ partial wave [10]. This raises the possibility of detecting the formation of positronic bound states by an optical ($^2P^o \rightarrow {}^2S^e$) transition between the e^+Ca states [11]. There is also the possibility of performing a far more challenging experiment using a low-energy Ps-beam scattering off Ca where $E_{\mathrm{Ps}} > 5.619$ eV produces charge transfer [10].

A large set of model CI calculations were undertaken to examine the systematics of excited states with $L = 1$. The Hamiltonian was chosen to be identical to that used previously for an investigation of positron binding to a model (one-electron) alkali atom [5], so only a brief summary of the technical details is given here. Firstly, the singly-charged atomic ionic core was based on the $1s^22s^22p^6$ orbitals taken from a Hartree-Fock calculation of neutral sodium. The exact core-valence exchange operator was replaced by a local exchange potential, tuned to reproduce the low-lying neutral sodium spectrum. Finally, an additional potential of the form $V_{\rm model}(r) = d_1 r \exp(-\frac{3}{2}r)$ is added where $d_1 \in [0.2, -5.6]$ a.u., giving ionisation energies $I.P. \in [4.8, 13.0]$ eV, and neutral atom dipole polarisabilities, $\alpha_d \in [209, 23.5]$ a_0^3.

The present CI calculations were performed with a minimum of 25 single-electron and 25 single-positron Laguerre-type orbitals for each partial wave [12]. The same sets of orbitals were then LS-coupled together such that the overall symmetry of the electron-positron wavefunction was either ^{1}S^e, or ^{1}P^o. Due to computational limitations, L_{max} was chosen to be 15 for ^{1}S^e while $L_{max} = 9$ was chosen for ^{1}P^o.

Figure 2 shows the binding energy of the ground state of the ^{1}S^e symmetry (the upper three curves), and the first-excited state in the ^{1}P^o symmetry (the lower two curves). The top-most curve is the set of previous fixed-core SVM (FCSVM) calculations [6], which provided close to converged ^{1}S^e values. The present CI calculations, denoted by the dashed-dot $L_{\rm max}$ curves, are seen to consistently lie below those values, particularly for the lower ionisation potentials. This is the region where the wavefunction is dominated by Ps, and the CI expansion is very slowly convergent with respect to $L_{\rm max}$ [13, 9]. The final two dotted curves in Figure 2, marked by $L = \infty$, are the results of performing a sequence of three smaller CI calculations with $L_{max} - 1$, $L_{max} - 2$, and $L_{max} - 3$, and then using known convergence patterns to extrapolate to the $L = \infty$ limit [14]. The resulting $L = \infty$ CI ^{1}S^e binding energies are significantly improved by this procedure, and visually are as good as the SVM for $I.P. > 6.0$ eV.

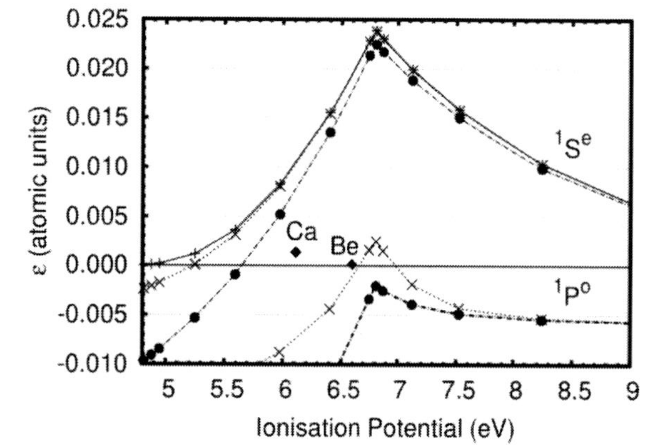

Figure 2. Binding energies of a model positronic alkali atom in the ^{1}S^e and ^{1}P^o symmetries. The (——) solid line is from FCSVM calculations of the ^{1}S^e series [5]. The two (— · —) dash-dot lines are from the CI $L_{\rm max}$ calculations, the two (· · · · · ·) dotted lines are the $L = \infty$ extrapolated results, the two (♦) results are from CI calculations [10], as discussed in the text.

The $L_{\rm max}$ ^{1}P^o calculations are all unbound, although a modest enlargement of the basis should produce binding. The extrapolation procedure to $L = \infty$ shows a small region where the system is actually bound. The 'stable' ionisation potential range is likely to be a minimum, since larger CI calculations will improve the description of a Ps-cluster orbiting the Atom$^+$ ion. There are no equivalent $L = 1$ FCSVM calculations to compare with.

There are no actual one-electron atom energies that lie in this range, however these calculations do provide additional motivation for the existance of the known two-electron excited positronic atoms. Shown for comparison in Figure 2 are the binding energies of the two known two-electron, one positron systems [10]. These correspond to the ^{2}P^o state of e^+Ca, and of a positron bound to metastable (^{3}P^o) Be. Essentially, the two-electron systems with low-ionisation energies have an additional p-wave binding mechanism, as they are expected to have a significant component of Ps$^-$, which will always bind in all partial waves to the residual Atom^{2+} [10].

Some other states were also investigated. Firstly, the second eigenstate in the ^{1}S^e symmetry was also computed as a by-product of the calculations shown in Figure 2. None of the excited state energies indicated binding, nor did their extrapolation. For example, for the model atom with $d_1 = -1.95$, (and an $I.P. = 6.8112$ eV) the ground state has the radius of the electron

$\langle r_e \rangle = 4.29$ a$_0$, while the positron has $\langle r_p \rangle = 6.67$ a$_0$. For the first excited state $\langle r_e \rangle = 3.37$ a$_0$, which is roughly that of the parent model atom, while $\langle r_p \rangle = 32.5$ a$_0$, which is heading out towards the extent of the Laguerre-based radial basis.

The unnatural parity, ^{1}P^e, state is also of fundamental interest. This is due to recent large scale CI calculations that showed the existence of a second, annihilation suppressed, unnatural parity, ^{2,4}S^o, state of PsH, NaPs and KPs [15, 16]. The positron is essentially bound to the second bound states of the H$^-$, Na$^-$ and K$^-$ ions which have np^2 ^{1}P^e-type configurations [17, 18]. This has sparked some interest in other systems such as the unnatural parity resonances in e^+-H [19].

The ^{2,4}S^o LiPs state, however, appears to be Borromean [16]. It is necessary to demonstrate that all of the parent ^{3}P^e states Li, e^+Li, and Ps are all themselves unstable. Firstly, both Li and Ps unnatural states are known to be unstable [18, 20]. Secondly, both CI and SVM calculations found no evidence for the stability of both Li and e^+Li unnatural states [16].

To support the conclusion that the ^{1}P^e state of e^+-Li is unstable, a series of ^{1}P^e model-alkali atoms were also computed with the CI basis as above. None of these states, across the entire range of ionisation potentials, showed any sign of binding. This is best seen in the $\langle r_p \rangle \approx 29.4$ a$_0$ across the entire range of d_1, while the $\langle r_e \rangle$ stays at roughly the radius of the parent atom. These results suggest that e^+Na and e^+K are also not stable in the ^{1}P^e symmetry.

4. Positron-atom scattering resonances

The e^+Mg ^{2}P^o system was used to introduce a method of quantum scattering that extracts phase shifts from pseudostate energy shifts [21], which essentially relies on the fact that a finite-L^2 basis forms a soft-box at some radius [22]. In contrast to the situation for e^+Ca, large-scale CI calculations have shown that e^+Mg has a low-lying ^{2}P^o shape resonance as opposed to a true bound state [21]. Using the pseudostate-shift method the elastic scattering cross sections were computed for e^+Mg, e^+Cu and e^+Zn [22]. These results are shown in Figure 3, along with preliminary calculations of e^+Cd. Also shown is the total cross section for e^+-Mg scattering as measured by the Detroit group, which had a beam width of approximately 1 eV. Energies of sub-100 meV are accessible with present positron-beam technologies [23].

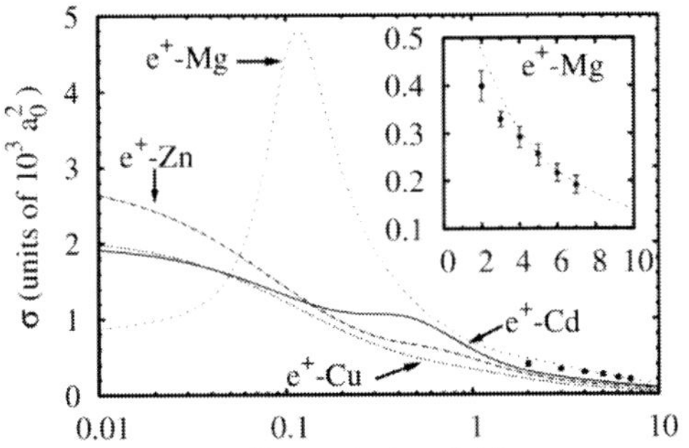

Figure 3. The elastic scattering cross section, σ (in units of 10^3 a$_0^3$), for e^+-Mg, e^+-Cu and e^+-Zn as a function of energy (in eV) [22]. Also shown are preliminary calculations of e^+-Cd scattering. The experimental (total) cross section for e^+-Mg scattering is also given [24].

Apart from the elastic cross section, it is straightforward with the pseudostate shift method to compute the annihilation parameter, $Z_{\rm eff}$ [22]. The results for e^+Mg, e^+Cu, e^+Zn and e^+-Cd are shown in Figure 4, where the two weak p-wave resonances are more prominent than in the elastic cross section. The systematic variation in the $Z_{\rm eff}$ resonance position and width between atoms was previously predicted with scattering from a simple exploratory-model potential [25], which supports the well-known idea that a p-wave resonance makes a transition to the bound state as the potential deepens, effectively pulling it under-water.

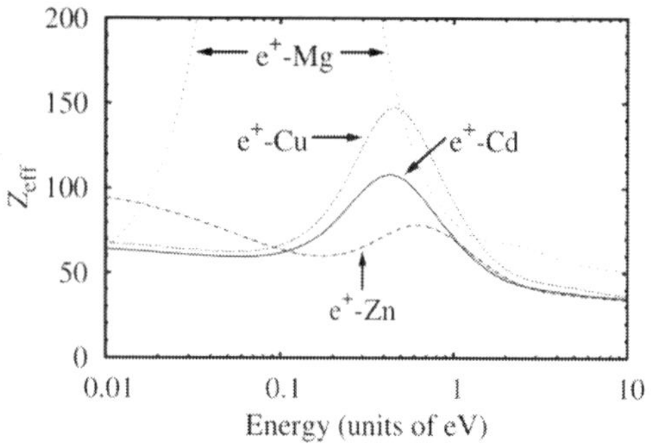

Figure 4. The positron annihilation parameter, $Z_{\rm eff}$, for e^+-Mg, e^+-Cu and e^+-Zn scattering as a function of energy (in eV) [22]. Also shown are preliminary calculations of e^+-Cd scattering.

5. Conclusions

Our understanding of positron-atom and positronium-atom excited states has seen rapid progress in the last few years. The calculations presented here of a simple model-alkali atom helps to explain the positron physics involving non s-wave states. The accurate description of these excited states with CI wavefunctions is extremely challenging due to the presence of positronium. The main result is that the ^{1}P$^{\rm o}$ symmetry does admit bound states over a limited range of atoms. This, however, should be merely taken as the first estimate. CI calculations with a extrapolation of the radial basis towards a complete basis set would be more definitive [26].

Acknowledgments

This work was partially supported under the Australian Research Council's Discovery Program (# 0665020). MB thanks Dr. James Otto of the SDSU Computational Science Research Center for system administration, and Dr. James Danielson and Prof. Cliff Surko for their interest.

References

[1] Mitroy J, Bromley M W J and Ryzhikh G G 2002 *J. Phys. B* **35** R81
[2] Gribakin G F, Young J A and Surko C M 2009 *Rev. Mod. Phys* under review
[3] Danielson J R, Young J A and Surko C M 2009 *Phys. Rev. A* under review
[4] Gribakin G F and Lee C M R 2006 *Nucl. Instrum. Methods Phys. Res. B* **247** 31
[5] Mitroy J, Bromley M W J and Ryzhikh G G 1999 *J. Phys. B* **32** 2203
[6] Mitroy J and Ryzhikh G G 1999 *J. Phys. B* **32** 1375
[7] Mitroy J 2000 *J. Phys. B* **33** 5307
[8] Mitroy J 2002 *Phys. Rev. A* **66** 010501(R)
[9] Bromley M W J and Mitroy J 2002 *Phys. Rev. A* **66** 062504
[10] Bromley M W J and Mitroy J 2007 *Phys. Rev. A* **75** 042506
[11] Bromley M W J and Mitroy J 2006 *Phys. Rev. Lett.* **97** 183402
[12] Bromley M W J, Mitroy J and Ryzhikh G G 2000 *Nucl. Instrum. Methods Phys. Res. B* **171** 47
[13] Mitroy J and Ryzhikh G G 1999 *J. Phys. B* **32** 2831
[14] Mitroy J and Bromley M W J 2006 *Phys. Rev. A* **73** 052712
[15] Mitroy J and Bromley M W J 2007 *Phys. Rev. Lett.* **98** 063401
[16] Bromley M W J, Mitroy J and Varga K 2007 *Phys. Rev. A* **75** 062505
[17] Drake G W F 1970 *Phys. Rev. Lett.* **24** 126
[18] Norcross D W 1974 *Phys. Rev. Lett.* **32** 192
[19] Yan Z C and Ho Y K 2008 *Phys. Rev. A* **77** 030701(R)
[20] Mills, Jr A P 1981 *Phys. Rev. A* **24** 3242
[21] Mitroy J and Bromley M W J 2007 *Phys. Rev. Lett.* **98** 173001
[22] Mitroy J, Zhang J Y, Bromley M W J and Young S I 2008 *Phys. Rev. A* **78** 012715
[23] Surko C M, Gribakin G F and Buckman S J 2005 *J. Phys. B* **38** R57
[24] Stein T S, Harte M, Jiang J, Kauppila W E, *et al.* 1998 *Nucl. Instrum. Methods Phys. Res. B* **143** 68
[25] Bromley M W J and Mitroy J 2003 *Phys. Rev. A* **67** 062709
[26] Bromley M W J and Mitroy J 2007 *Int. J. Quantum Chem.* **107** 1150

Analysis of experimental positron-molecule binding energies

J. R. Danielson, J. A. Young[1], and C. M. Surko

Department of Physics, University of California, San Diego La Jolla, CA 92093-0319, USA

E-mail: `csurko@ucsd.edu`

Abstract. Experiments show that positron annihilation on molecules frequently occurs via capture into vibrational Feshbach resonances. In these cases, the downshifts in the annihilation spectra from the vibrational mode spectra provide measures of the positron-molecule binding energies. An analysis of these binding energy data is presented in terms of the molecular dipole polarizability, the permanent dipole moment, and the number of π bonds in aromatic molecules. The results of this analysis are in reasonably good agreement with other information about positron-molecule bound states. Predictions for other targets and promising candidate molecules for further investigation are discussed.

1. Introduction

While positrons are of use in many areas of science and technology, a quantitative understanding of their behaviour in many physical situations is lacking [1, 2, 3]. One example is the binding of positrons to ordinary matter (e.g., atoms and molecules). There are calculations that are believed to be quite accurate for the binding of positrons to atoms, however there are as yet no experimental measurements of positron-atom binding energies [4, 5, 6]. The situation is essentially the reverse in the case of positron binding to molecules. Binding energies have now been measured for thirty molecules [7, 8], but few calculations are available and those that have been done are generally for molecules with large permanent dipole moments.

Here, we conduct an analysis to express the positron-molecule binding energy, $\epsilon_{\rm b}$, in terms of physical properties of the molecule including the dipole polarizability, the permanent dipole moment, and the number of π bonds in aromatic molecules. The predictions of this analysis are tested with other available data. They are also used to make a number of predictions, including the identification of molecules with unusually large binding energies and interesting chemical trends and candidates for further study.

2. Measurements of positron-molecule binding energies

Annihilation rates on atoms and molecules are usually expressed as a normalized rate, $Z_{\rm eff}$, which is the measured annihilation rate, λ, normalized to that of a free electron gas [9], specifically

$$Z_{\rm eff} = \frac{\lambda}{\pi r_0^2 c n_{\rm m}}, \tag{1}$$

[1] Currently at Jet Propulsion Laboratory, Pasadena, CA

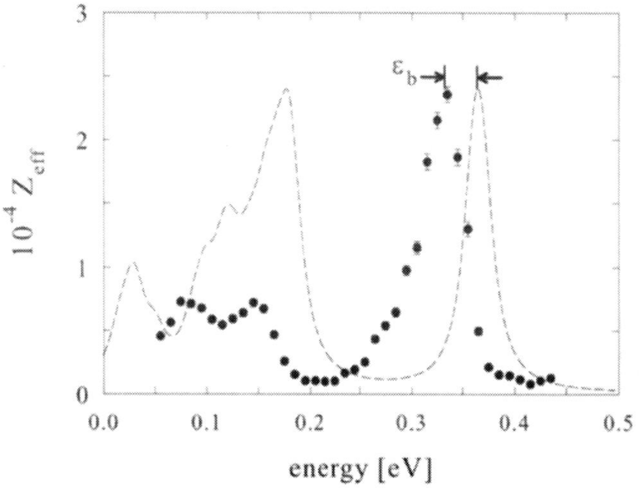

Figure 1. The normalized annihilation spectrum, $Z_{\rm eff}(\epsilon)$, for butane (C_4H_{10}) is shown as a function of incident positron energy. Also shown (- -) is the spectrum of vibrational modes, in arbitrary amplitude units, with each mode arbitrarily broadened in energy by 10 meV FWHM. The shift of $\sim$ 35 meV between the two spectra is a measure of the positron-binding energy, $\epsilon_{\rm b}$.

where r_0 is the classical electron radius, c is the speed of light, and $n_{\rm m}$ is the number density of the target atoms or molecules. In this representation, the value $Z_{\rm eff} = Z$, where Z is the number of electrons in the target, corresponds to annihilation in an uncorrelated electron gas with electron density $n_{\rm m}Z$.

High-quality, trap-based positron beams have enabled measurements of $Z_{\rm eff}(\epsilon)$ as a function of the incident positron energy, ϵ [10]. The techniques used for these measurements are described, for example, in Refs. [7, 8]. They use a tuneable, cold positron beam magnetically guided through a gas cell. The result is an asymmetric distribution in total positron energy with a width $\sim$ 40 meV FWHM [7, 11]. An example of such a measurement is shown in Fig. 1. It is typical of the spectra for many molecules in that $Z_{\rm eff}(\epsilon)$ shows evidence of vibrational Feshbach resonances (VFR). In this resonant annihilation process, the incident positron excites a molecular vibration and makes a transition from the continuum into a positron-molecule attached state [10, 12]. This process *requires* the existence of a positron bound state on the molecule. The observed energy of the resonance, $\epsilon_{\rm res}$, is determined by energy conservation, namely

$$\epsilon_{\rm res} + \epsilon_{\rm b} = \epsilon_{\nu}, \tag{2}$$

where ϵ_ν is the energy of the molecular vibration. Hence the downshift, $\Delta\epsilon$, of the resonance with respect to ϵ_ν is a direct measure of the binding energy [11]. The attached state is a resonance and not a true bound state in that the vibrational energy in the molecule can subsequently eject the positron.

Studies of resonances from different vibrational modes and from deuterated hydrocarbons indicate that the presence of the positron does not alter the vibrational mode energy [7]. Thus the binding energies measured using this technique are valid measures of the binding energy for the ground-state molecule.

The measurements of $\epsilon_{\rm b}$ obtained from $Z_{\rm eff}(\epsilon)$ spectra for smaller molecules are taken from Ref. [8] and for larger molecules from Ref. [7]. The absolute accuracies of the binding energy measurements for those in which the shifts were resolved directly are typically $\sim$ 10 - 15 meV. A summary of data for molecules used in the analysis presented here are shown in Table I, together with the molecular parameters relevant to the analysis.

3. Fit to molecular parameters

The goal of the present analysis is to find a simple parameterization of the binding energy in terms of molecular properties. Two effects that can give rise to binding include the dipole

Table 1. The values of binding energy, ϵ_b [meV], are tabulated for all molecules where ϵ_b has been measured [7, 8], along with the predictions of Eq. (3). Also listed are the number of electrons, Z, on the molecule; dipole polarizability, α [Å^3] [13, 14]; permanent dipole moment μ [D] [13, 15]; and the ionization energy, E_i [eV] [13]. Values in parentheses are lower bounds for molecules that lack known values for μ.

Molecule Name	Formula	ϵ_b [meV]	Eq. (3) [meV]	Z	α [Å^3]	μ [D]	E_i [eV]
Alkanes							
Ethane	C_2H_6	≥ 0	-14	18	4.4	0	11.5
Propane	C_3H_8	10	10	26	6.3	0.1	11.1
Butane	C_4H_{10}	35	31	34	8.1	0	10.6
Pentane	C_5H_{12}	60	54	42	10.0	0	10.4
Hexane	C_6H_{14}	80	77	50	11.8	0	10.2
Heptane	C_7H_{16}	105	100	58	13.7	0	9.9
Octane	C_8H_{18}	115	123	66	15.5	0	10.0
Nonane	C_9H_{20}	145	146	74	17.4	0	10.0
Dodecane	$C_{12}H_{26}$	220	214	98	22.9	0	9.9
Tetradecane	$C_{14}H_{30}$	260	261	114	26.6	0	—
Hexadecane	$C_{16}H_{34}$	310	306	130	30.3	0	9.9
Deuterated alkanes							
Butane-d10	C_4D_{10}	35	31	34	8.1	0	—
Nonane-d20	C_9D_{20}	145	146	74	17.4	0	—
Alkane related molecules							
Acetylene	C_2H_2	≥ 0	-28	14	3.3	0	11.4
Ethylene	C_2H_4	20	-17	16	4.2	0	10.5
Isopentane	C_5H_{12}	60	57	42	10.0	0.1	10.3
Cyclopropane	C_3H_6	10	1	24	5.7	0	9.9
Cyclohexane	C_6H_{12}	80	68	48	11.1	0	9.9
Aromatics [N_π term included in Eq. (5)]							
Benzene	C_6H_6	150	149	42	10.4	0	9.3
Benzene-d6	C_6D_6	150	149	42	10.4	0	9.3
Naphthalene	$C_{10}H_8$	300	296	68	16.6	0	8.2
Alcohols							
Methanol	CH_3OH	2	5	18	3.3	1.7	10.9
Ethanol	C_2H_5OH	45	27	26	5.1	1.7	10.5
Partially halogenated hydrocarbons							
Methyl fluoride	CH_3F	≥ 0	-3	18	2.4	1.85	12.9
Methyl chloride	CH_3Cl	25	23	26	4.4	1.9	11.2
Methyl bromide	CH_3Br	40	35	44	5.6	1.8	10.5
1-fluoropropane	C_3H_7F	35	45	34	6.0	2.0	11.3
2,2-difluoropropane	$C_3H_6F_2$	25	51	42	5.9	2.4	11.4
1-fluorobutane	C_4H_9F	70	(27)	42	7.8	–	—
1-fluorohexane	$C_6H_{13}F$	80	(73)	58	11.5	–	—
1-fluorononane	$C_9H_{19}F$	145	(141)	82	17.0	–	—
1-chlorohexane	$C_6H_{13}Cl$	175	138	66	13.6	2.0	10.3

polarizability and the permanent dipole moment of the molecule. It turns out that this will leave an unexplained deficit in binding energy for aromatic molecules, and so an additional parameter, the number of molecular π bonds in aromatic molecules, is also included.

We have made extensive studies of alkane molecules (C_nH_{2n+2}). They represent a case in which the molecules do not have either a permanent dipole moment or π bonds. Further, the molecular size (and hence polarizability) of these molecules can be varied without changing the chemical structure. Thus we fix the dependence upon the molecular polarizability by fitting to these molecules. As shown in Fig. 2, the alkane binding energies are, to good accuracy, linear in the molecular polarizability, α. To accommodate molecules with permanent dipole moments,

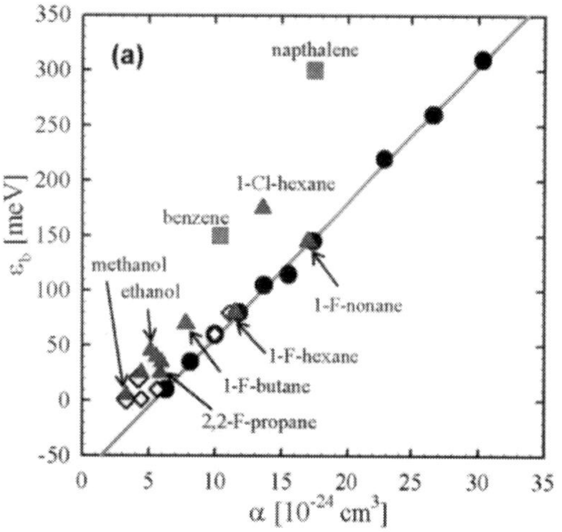

Figure 2. Measured positron binding energies, ϵ_b, as a function of the molecular dipole polarizability, α. Alkane molecules used in fit (•); alkane related molecules (◇); molecules with permanent dipole moments (▲); and aromatics with π bonds (■).

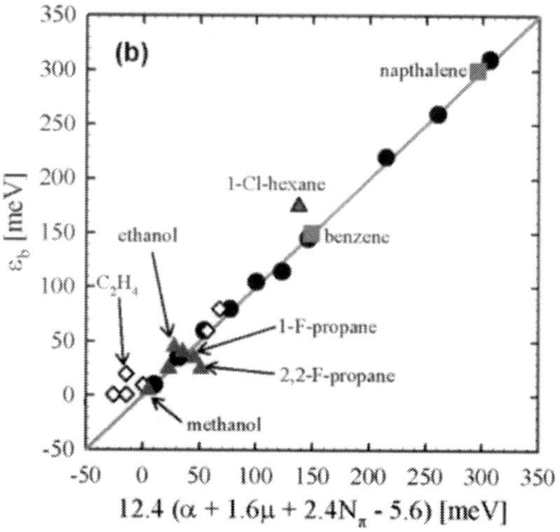

Figure 3. Binding energy fit as described by Eq. (3), using the polarizability α, permanent dipole moment, μ, and the number of π bonds, N_π, for aromatic molecules.

an additive term linear in μ is included to model its effect on ϵ_b. In this paper, α is expressed in units of 10^{-24} cm^{-3}, and the permanent dipole moment, μ, is expressed in units of debye (D).

As mentioned above, the fit to α and μ leaves a significant discrepancy for the two aromatic molecules in the data set, benzene and naphthalene. We speculate that double bonds involving π orbitals can attract the positron more strongly, since the electron density in these bonds is farther from the (repulsive) cores of the carbon atoms. For these aromatics, the number of π bonds, N_π, turns out to be a convenient parameter ($N_\pi = 3$ for benzene and 5 for naphthalene).

The fit to alpha for alkanes was augmented by terms linear in μ, and N_π (i.e., the latter for the two aromatic molecules only) while the coefficients for the slope and offset from the alkane fit are held constant. The result is

$$\epsilon_b = 12.4(\alpha + 1.6\mu + 2.4N_\pi - 5.6) \ \text{[meV]}. \tag{3}$$

This fit to the binding energy data is shown in Fig. 3, with the numerical values given in Table I. As indicted in Fig. 3 and Table I, the binding energies for most molecules are in reasonably good agreement with the predictions of Eq. (3). The most notable exceptions are 2,2 difluoropropane, ethylene and 1-chlorohexane.

4. Discussion

A number of checks on the model can be made. For example, studies of a number of small molecules indicate the absence of VFR and this is interpreted as evidence that these molecules do not bind positrons. Generally, Eq. (3) predicts that this is the case. For example, H_2, O_2, CO, NO, CO_2, CH_4 and CF_4 do not exhibit VFR. They are predicted to have $\epsilon_b < -30$ meV, which is consistent with the model. As can be seen in Table I, disagreements include acetylene and ethylene.

Calculations for binding to molecules have been conducted for several molecules, all of which have appreciable dipole moments. The predictions from these theoretical calculations are listed

Journal of Physics: Conference Series **199** (2010) 012012 doi:10.1088/1742-6596/199/1/012012

Table 2. Comparisons of theoretical predictions for positron-molecule binding energies with those from Eq. (3).

Name	Formula	α [Å^3]	μ [D]	ϵ_b, Eq. (3) [meV]	ϵ_b, th [meV]
Cyanide	HCN	2.5	3.0	21	35[a]
Formaldehyde	H_2CO	2.8	2.3	11	18[b]
Urea	$(NH_2)_2CO$	9.7	4.6	140	13[c]
Acetone	$(CH_3)_2CO$	6.4	2.9	67	4[c]
Lith. hydride	LiH	3.8	5.9	95	1000[d]

[a]Ref. [17]. [b]Ref. [16]. [c]Ref. [18]. [d]Refs. [19, 20, 21, 22].

in Table II, and they are also in reasonable agreement with those from Eq. (3). Cyanide and formaldehyde agree to within a factor of two. While the theoretical values for urea and acetone are about an order of magnitude smaller than those from Eq. (3), it is believed that these calculations likely underestimate ϵ_b [16], in agreement with the present comparison.

The predictions of Eq. (3) can also be compared with electron-molecule attached states (i.e., negative ions). There are two classes of negative molecular ions. When the molecule has an open valence shell, the electron forms a strong bond with the molecule, and this typically results in electron-volt binding energies; these states have no positron analogue. If, on the other hand, the valence shell is closed, the electron can be weakly bound to the molecule due to the forces associated with the dipole polarizability and/or the permanent dipole moment of the molecule. These latter states approximate closely the positron-attached states described here. In fact, (spherical) atomic clusters of CO_2 have been shown to exhibit VFR, and the spectra of the resulting negative ions that are formed as a function of incident electron energy have been used to measure the electron binding energies [23]. This is very similar to the procedure for the positron-molecule VFR measurements described here. Further, the binding energy is also found to be an approximately linear function of atomic polarizability, consistent with Eq. (3), namely $\epsilon_b \approx 4.0(\alpha - 7.5)$ [meV]. It is interesting that the coefficient of proportionality $\Delta\epsilon_b/\Delta\alpha$ is a factor of three smaller in the case of electrons. This appears to be an important clue that the linear dependence of ϵ_b reflects more than simply long-range electrostatic coupling (i.e., which would be the same for electrons and positrons). In the positron case, it is possible that the (closed) positronium channel plays a role in enhancing the binding of the positron to the target.

Another point of comparison between electron- and positron-attached states is the case of molecules with permanent dipole moments. Again, in the case of molecules with closed valence shells, a permanent dipole moment on a molecule is found to produce a weakly bound negative ion. Examples of such negative ion states include acetone (μ = 2.9; ϵ_b = 1.5 meV) and acetonitrile (μ = 3.9; ϵ_b = 11.5 meV) [24]. Theoretically, a permanent moment of 1.625 D will bind an electron or positron [25]. Experimentally, to bind an electron with greater than 1 meV binding energy requires a moment in excess of $\sim$ 2.5 D [24]. This experimental result is in reasonable agreement with the critical value for positrons of 3.5 D obtained from Eq. (3).

Equation (3) can also be used to make a number of interesting predictions. Large targets typically have large dipole polarizabilities and hence they will generically tend to bind positrons. Examples include the larger fluoroalkanes such as fluorohexane (α = 12.4 cm^{-3}, ϵ_b = 84) and C_{60}(α = 76 cm^{-3}, ϵ_b = 880). In this regard, if ϵ_b is greater than the most energetic fundamental mode frequency, then the primary VFR will be shut off. This has been observed for the case of fully deuterated naphthelene [7]. The cluster C_{60} is also expected to be such a case.

5. Concluding remarks

We describe here an empirical parameterization of positron-molecule binding energies in terms of molecular parameters. It indicates that the dominant contributions come from the molecular dipole polarizability and permanent dipole moment, augmented by an additional term for aromatic molecules. In large molecules, the first and last terms dominate. The positron-molecule complexes discussed here have close analogues of a certain class of weakly bound electron-molecule attached states (i.e., molecules with closed valence shells). One interesting feature of this comparison is the fact that the polarizability term appears to be much stronger in the positron case.

The parameterization in Eq. (3) is expected to be useful in identifying targets for future study. One important example is small molecules that appear to have appreciable binding energies and can both be studied experimentally and calculated using currently available theoretical tools. Other interesting targets include large perfluoroalkanes, larger polycyclic aromatic molecules such as anthracene and pyrene, and atomic clusters such as C_{60}. A more detailed description of this analysis and its implications, including the relationship between the predictions of Eq. (3) and the extensive body of data for molecular annihilation rates using thermal positrons at 300 K, will be published elsewhere [26].

Acknowledgments
We wish to thank G. Gribakin and M. Bromley for helpful conversations. This work is supported by the U. S. NSF, grant PHY 07-55809.

References
[1] M. Charlton and J. W. Humberston, Positron Physics (Cambridge Univ. Press, Cambridge, U. K., 2001).
[2] P. G. Coleman, Positron Beams and Their Applications (World Scientific, Singapore, 2000).
[3] C. M. Surko, G. F. Gribakin, and S. J. Buckman, J. Phys. B: At. Mol. Opt. Phys. **38**, R57 (2005).
[4] J. Mitroy, M. W. J. Bromley, and G. G. Ryzhikh, J. Phys. B: At. Mol. Opt. Phys. **35**, R81 (2002).
[5] G. G. Ryzhikh, J. Mitroy, and K. Varga, J. Phys. B. **31**, 3965 (1998).
[6] D. Bressanini, M. Mella, and G. Morosi, J. Chem. Phys. **108**, 4756 (1998).
[7] J. A. Young and C. M. Surko, Phys. Rev. A **77**, 052704 (2008).
[8] J. A. Young and C. M. Surko, Phys. Rev. A **78**, 032702 (2008).
[9] P. A. Fraser, Adv. in At. and Mol. Phys. **4**, 63 (1968).
[10] S. J. Gilbert, L. D. Barnes, J. P. Sullivan, and C. M. Surko, Phys. Rev. Lett. **88**, 043201 (2002).
[11] G. F. Gribakin and C. M. R. Lee, Phys. Rev. Lett. **97**, 193201 (2006).
[12] L. D. Barnes, S. J. Gilbert, and C. M. Surko, Phys. Rev. A **67**, 032706 (2003).
[13] *CRC Handbook of Chemistry and Physics*, 89th Ed., 2008/2009, editor David R. Lide, CRC Press/Taylor and Francis, Boca Raton, FL.
[14] K. J. Miller, J. Am. Chem. Soc. **112**, 8533 (1990).
[15] A.L. McClellan, *Tables of experimental dipole moments*, W.H. Freeman and Company, San Francisco, 1963.
[16] K. Strasburger, Struct. Chem. **15**, 415 (2004).
[17] H. Chojnacki and K. Strasburger, Mol. Phys. **104**, 2273 (2006).
[18] M. Tachikawa, R. J. Buenker, and M. Kimura, J. Chem. Phys. **119**, 5005 (2003).
[19] R. J. Buenker, H. P. Liebermann, V. Melnikov, M. Tachikawa, L. Pichl, and M. Kimura, J. Phys. Chem. A **109**, 5956 (2005).
[20] K. Strasburger, J. Chem. Phys. **114**, 615 (2001).
[21] R. J. Buenker, H. P. Liebermann, M. Tachikawa, L. Pichl, and M. Kimura, Nucl. Instrum. and Meth. B **247**, 47 (2006).
[22] M. Mella, G. Morosi, D. Bressanini, and E. Stefano, J. Chem. Phys. **113**, 6154 (2000).
[23] E. Leber, S. Barsotti, I.I. Fabrikant, J.M. Weber, M.-W. Ruf, and H. Hotop, Eur. Phys. J. D **12**, 125 (2000).
[24] C. Desfrancois, H. Abdoul-Carime, N. Khelifa, and J. P. Schermann, Phys. Rev. Lett. **73**, 2436 (1994).
[25] E. Fermi and E. Teller, Phys. Rev. **72**, 399 (1947); J. M. Levy-Leblond, Phys. Rev. **153**, 1 (1967).
[26] J. R. Danielson, J. A. Young, and C. M. Surko, J. Phys. B: At. Mol. Opt. Phys. **42**, 235303 (2009).

Journal of Physics: Conference Series **199** (2010) 012013 doi:10.1088/1742-6596/199/1/012013

Resonant positron annihilation in ammonia

G. F. Gribakin

Department of Applied Mathematics and Theoretical Physics, Queen's University, Belfast BT7 1NN, Northern Ireland, UK

E-mail: g.gribakin@qub.ac.uk

Abstract. Positron annihilation in ammonia is analyzed using the framework of resonant annihilation [G. F. Gribakin and C. M. R. Lee, Phys. Rev. Lett. **97**, 193201 (2006)]. In particular, we show that molecular rotations can have a measurable effect on the annihilation rates at room temperatures. Rotation leads to broadening of vibrational Feshbach resonances. Rotations also allow a distinct contribution at low positron energies in the form of a rotational Feshbach resonance. This resonance can enhance the annihilation rate for thermalized room-temperature positrons. Comparison of theory and experiment shows that overtone and combination vibrations, including those due to inversion doubling, likely play an important role.

1. Introduction

The purpose of this work is to analyze the role of resonances in the positron annihilation in ammonia (NH_3). In particular, we estimate the effect of molecular rotations in broadening the observed vibrational resonances. A new type of resonance which can occur at very low positron energies for molecules with nonzero permanent dipole moments is also identified. This *rotational Feshbach resonance* can have a sizeable affect on the annihilation of thermal positrons in small polar molecules.

Resonant processes have long been suspected to be the origin of anomalously large positron annihilation rates observed in many polyatomic gases [1, 2, 3, 4]. This explanation remained a hypothesis until vibrational resonances were actually observed using a high-resolution trap-based positron beam [5]. Since then, this technique has been used to measure the annihilation rates as a function of the positron energy for a large number of large and small polyatomic molecules [6, 7, 8, 9].

In general, vibrational Feshbach resonances (VFR) can be observed in molecules capable of binding the positron. They occur when an incident positron with energy $\varepsilon > 0$ is captured into the bound state at negative energy $-\varepsilon_b$, the energy $\varepsilon + \varepsilon_b$ being transferred into the vibrational motion of the positron-molecule complex. The energy conservation for this process determines the energies of the resonances,

$$\varepsilon_\nu = E_\nu - \varepsilon_b, \tag{1}$$

where ε_b is the positron-molecule binding energy, and E_ν is the energy of the molecular vibrational excitation. According to Eq. (1), the VFRs are downshifted relative to the vibrational excitation energies by the positron binding energy. Observation of resonances thus allows one to measure ε_b. At present, positron binding energies for about 30 molecules have been determined in this way [10].

Theoretically, resonant annihilation can be described using the simple Breit-Wigner formalism [3, 4]. Its application, however, requires knowledge of the positron-molecule binding energy, vibrational excitation energies of the positron-molecule complex (including combination vibrations and overtones), as well as the total and partial widths of the resonances, i.e., those for annihilation and positron detachment. So far, it has been possible to calculate the annihilation rates only for small-sized polyatomic molecules with infrared-active modes, such as protonated and deuterated methyl halides [11, 12]. The binding energy is a free parameter in the theory. It is determined by comparison with the energy-resolved experimental annihilation rates. Besides methyl halides, the theory also provided a good description of the experiment in ethanol [9] and methanol, in which a contribution from overtones and combination vibrations had to be included [12]. Its application to small polyatomics in which not all modes are infrared active, e.g., acetylene and ethylene, indicates that VFR of infrared-inactive modes, as well as overtones and combinations, need to be considered to explain the experimental data.

Ammonia is the smallest molecule with infrared-active modes whose annihilation rate has been measured [9, 13]. It does not show the vibrational resonances as clearly as the molecules mentioned above. However, the magnitude of the annihilation rate suggests that it contains a large contribution of resonant annihilation. In this paper we use the theory of Ref. [11] and its extension in order to understand the details of positron annihilation in NH_3.

2. Resonant annihilation

The positron annihilation cross section for many-electron targets is usually parameterized in terms of the effective number of electrons $Z_{\rm eff}$,

$$\sigma_a = \pi r_0^2 \frac{c}{v} Z_{\rm eff}, \tag{2}$$

where r_0 is the classical electron radius, c is the speed of light, and v is the positron velocity. The *resonant* contribution to the annihilation cross section can be estimated using the Breit-Wigner theory [3, 4], which yields

$$Z_{\rm eff}^{\rm (res)} = \frac{\pi}{k}\rho_{ep}\sum_\nu \frac{g_\nu \Gamma_\nu^e}{(\varepsilon-\varepsilon_\nu)^2+\Gamma_\nu^2/4}. \tag{3}$$

Here Γ_ν^e and Γ_ν are the elastic and total widths of νth resonance, g_ν is its degeneracy, k is the positron momentum and ρ_{ep} is the average electron density at the positron in the bound state (atomic units are used throughout). This density determines the annihilation width of the resonances,

$$\Gamma^a = \pi r_0^2 c \rho_{ep}. \tag{4}$$

It scales with the binding energy as $\rho_{ep} = F\kappa/2\pi$, where $\kappa = \sqrt{2\varepsilon_b}$, and $F \approx 0.66$ [4]. The same parameters determine the contribution of *direct* annihilation, $Z_{\rm eff}^{\rm (dir)} \simeq F/(\kappa^2+k^2)$ [3].

To compare with experiment, the normalized annihilation rate, $Z_{\rm eff}$, must be averaged over the positron beam energy distribution. For the resonant part, this can be done analytically [11],

$$\bar{Z}_{\rm eff}^{\rm (res)}(\epsilon) = 2\pi^2\rho_{ep}\sum_\nu \frac{g_\nu\Gamma_\nu^e}{k_\nu\Gamma_\nu}\Delta(\epsilon-\varepsilon_\nu), \tag{5}$$

where $k_\nu = \sqrt{2\varepsilon_\nu}$, and the function $\Delta(E)$ describes the shape of a narrow resonance (e.g., $\Gamma_\nu \lesssim 1$ meV) observed with the trap-based positron beam (note that a factor 1/2 is missing in Eq. (10) of Ref. [11]),

$$\Delta(E) = \frac{1}{2k_BT_\perp}\exp\left[\frac{\sigma^2}{2(k_BT_\perp)^2}\right]\exp\left(\frac{E}{k_BT_\perp}\right)\left\{1+\Phi\left[-\frac{1}{\sqrt{2}}\left(\frac{E}{\sigma}+\frac{\sigma}{k_BT_\perp}\right)\right]\right\}. \tag{6}$$

Here $T_\perp$ is the transverse temperature of the positron beam, $\sigma = \delta_z/\sqrt{8\ln 2}$, δ_z being the full width at half maximum of the longitudinal energy distribution, k_B is Boltzmann's constant, and $\Phi(x)$ is the error function. The typical experimental values are $k_B T_\perp = \delta_z = 25$ meV.

The elastic resonance widths of infrared-active vibrational excitations can be estimated using their transition dipole amplitudes d_ν and frequencies ω_ν [11],

$$\Gamma_\nu^e = \frac{16\omega_\nu d_\nu^2}{27} h(\xi), \tag{7}$$

where $h(\xi) = \xi^{3/2}(1-\xi)^{-1/2}\left[{}_2F_1\left(\frac{1}{2},1;\frac{5}{2};-\xi/(1-\xi)\right)\right]^2$ is a dimensionless function of $\xi = 1-\varepsilon_b/\omega_\nu$. Owing to the weakness of positron binding ($\varepsilon_b \sim 1$–10 meV for small polyatomics), one can also assume that the vibrational excitation energies of the positron-molecule complex are close to those of the neutral molecule, $E_\nu \approx \omega_\nu$. The values of ω_ν of d_ν for many molecules are known from infrared absorption measurements [14].

The elastic width of a typical infrared-active VFR, $\Gamma_\nu^e \lesssim 0.1$ meV, is much larger than its annihilation width, e.g., $\Gamma^a = 0.1$ μeV for $\varepsilon_b = 10$ meV. The total width Γ_ν is then dominated by the elastic width (unless positron detachment accompanied by vibrational excitation of the molecule, i.e., vibrationally inelastic escape, is important). In this case the ratio Γ_ν^e/Γ_ν in Eq. (5) is close to unity, and the contribution of each resonance for a given molecule is determined by the factor g_ν/k_ν. In fact, one can also use Eq. (5) with $\Gamma_\nu^e/\Gamma_\nu = 1$ to estimate the contribution of VFR involving infrared-inactive modes. For these modes the long-range dipole interaction which results in Eq. (7), is zero, so the positron coupling to these modes is probably weaker. However, because of the smallness of Γ^a, they may still satisfy $\Gamma_\nu^e > \Gamma^a$, making $\Gamma_\nu^e/\Gamma_\nu \approx 1$ a reasonable assumption.

The NH_3 molecule has four vibrational modes with frequencies $\omega_\nu = 414$, 118, 427 and 201 meV, dipole amplitudes $d_\nu = 0.0095$, 0.0981, 0.0054 and 0.0227 a.u., and degeneracies $g_\nu = 1$, 1, 2 and 2, for the modes 1 to 4, respectively [14]. Using these values and assuming a small binding energy of $\varepsilon_b = 2$ meV, we obtain $\bar{Z}_{\rm eff}^{\rm (res)}$ from Eq. (5). It is shown in Fig. 1 together with the positron-energy averaged $Z_{\rm eff}^{\rm (dir)}$. One can immediately see that the theoretical resonances are much sharper than any features of the experimental $Z_{\rm eff}$ spectrum. The choice of the binding energy used is rather arbitrary. Apart from a small energy shift described by Eq. (1), the binding energy affects the overall magnitude of $Z_{\rm eff}^{\rm (res)}$ through the factor $\rho_{ep} \propto \sqrt{\varepsilon_b}$. Hence, the value chosen is a compromise between describing the features of the experimental data at low (0.1–0.2 eV) and high (0.35–0.45 eV) energies. Theory also fails to account for the signal between the resonance groups. This situation is in stark contrast to that in methyl halides, where theory and experiment are in good agreement [11, 12]. In the following sections we analyze effects which may account for this discrepancy.

3. Rotational broadening of VFR

In all previous estimates of resonant annihilation, rotations have been neglected. The positron-molecule bound states is likely to be the fully symmetric, i.e., A_1 or A_g. For smaller molecules and weak binding it is largely of s-wave character, behaving like a bound state in a zero-range potential $\varphi(\mathbf{r}) \simeq Ar^{-1}e^{-\kappa r}$ at large positron-molecule separations r. The incident positron wave function at low energies is also dominated by the s-wave component. In this case the positron capture into a VFR does not change the total angular momentum J of the molecule. This situation is realized in the zero-range potential model of positron annihilation with a dimer [15, 16].

For VFRs of infrared-active vibrations, the positron is captured into the symmetric bound state from the incident p wave. The selection rule for the vibrational and rotational transition is the same as for infrared absorption. For a symmetric top molecule, such as ammonia, this means

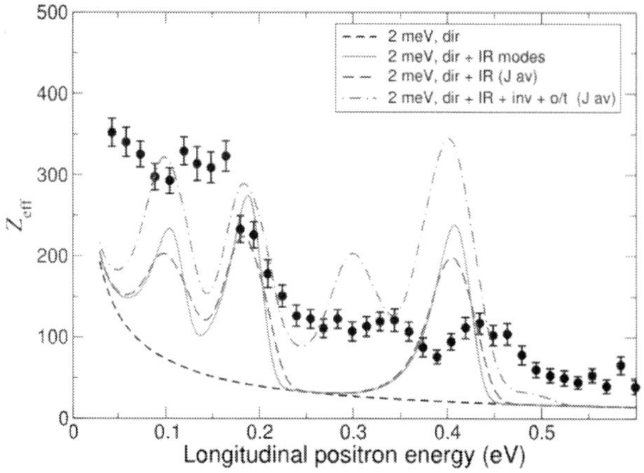

Figure 1. Normalized positron annihilation rate $Z_{\rm eff}$ in ammonia. Experimental data (●) obtained with the trap-based positron beam [13] (see also [9]). Theory (for $\varepsilon_b = 2$ meV): – – –, $Z_{\rm eff}^{\rm (dir)}$; ——, total $Z_{\rm eff} = Z_{\rm eff}^{\rm (dir)} + Z_{\rm eff}^{\rm (res)}$, including only mode-based VFR; — — —, total $Z_{\rm eff}$ with the resonances broadened by rotations at 300 K; — · —, total $Z_{\rm eff}$ including VFR due to the modes, inversion doubling and combination vibrations (rotationally broadened) .

$\Delta J = 0, \pm 1$, and $\Delta K = 0$ or $\Delta K = \pm 1$, depending on whether the transition moment is parallel or perpendicular to the molecular axis [17]. Here K is the quantum number of the component of the angular momentum along the molecular axis. Together, the J and K determine the rotational energy,

$$E_{JK} = BJ(J+1) + (C-B)K^2, \tag{8}$$

where $B = 1/2I_B$ and $C = 1/2I_C$, I_B and I_C being the principal moments of inertia perpendicular and parallel to the molecular axis. For ammonia the rotational constants are $B = 1.23$ meV and $C = 0.77$ meV [18].

In a dipole transition the rotational quantum numbers and the rotational energies of the initial and final state are different. This means that in an ensemble of molecules with different J and K, a number of resonances will appear near the energy ε_ν. Their statistical weights will depend on the Boltzmann factors $\exp(-E_{JK}/k_BT)$, where T is the molecular temperature. In current room-temperature experiments the energy spread of the positron beam is larger than the rotational energy shifts ΔE_{JK}. Hence, the energy shifts will lead to additional broadening of the observed resonances.

For a simple of estimate of this effect, we replace the symmetric top with a spherical top, characterized by the single average rotational constant $\bar{B}$, $3/\bar{B} = 2/B + 1/C$, i.e., $\bar{B} = 1.02$ meV for NH_3. Its rotational energy levels,

$$E_J = \bar{B}J(J+1) \tag{9}$$

are $(2J+1)^2$ degenerate, with populations $w_J = Z^{-1}(2J+1)^2 \exp(-E_J/k_BT)$, where

$$Z = \sum_J (2J+1)^2 \exp(-E_J/k_BT). \tag{10}$$

For $\Delta J = \pm 1$ transition, the rotational energy shifts are $\Delta E_{J+1,J} = 2\bar{B}(J+1)$. Adding these shifts to the resonance energies ε_ν, we average $Z_{\rm eff}^{\rm (res)}$ due to mode-based VFR over J with probabilities w_J. The result is shown by the long-dashed curve in Fig. 1. For the most probable J value, $J \sim \sqrt{2k_BT/\bar{B}} \sim 7$ at $T = 300$ K, the rotational broadening is $\Delta E_{J+1,J} \sim 2\sqrt{2k_BT\bar{B}} \sim 15$ meV. This broadening of the VFRs is indeed noticeable in Fig. 1. However, the calculated resonances remain sharper than the measured $Z_{\rm eff}$.

4. Rotational Feshbach resonance in NH_3

Another possible effect in ammonia is due to its the permanent electric dipole moment, $\mu = 1.47$ debye $= 0.578$ a.u. [19]. The positron interaction with the dipole, $V = \boldsymbol{\mu}\cdot\mathbf{r}/r^3$, can lead to positron capture accompanied by a *rotational* $J \to J+1$ excitation of the molecule. This resonance capture will occur at the positron energy

$$\varepsilon_J \equiv k_J^2/2 = 2\bar{B}(J+1) - \varepsilon_b. \tag{11}$$

The elastic width of the corresponding rotational Feshbach resonance (RFR) can be estimated similarly to that of the VFR of infrared-active modes [11],

$$\Gamma_J^e = \frac{8k_J^3}{27\kappa}\frac{J+1}{2J+1}\left|{}_2F_1\left(\tfrac{1}{2},1;\tfrac{5}{2};-\varepsilon_J/\varepsilon_b\right)\right|^2, \tag{12}$$

cf. Eq. (7). An estimate of this width shows that $\Gamma_J^e \gg \Gamma^a$, which means that the RFR will contribute to positron annihilation in the same way as any VFR. Note that RFRs were considered earlier in the context of low-energy electron collisions with simple molecules [20].

For NH_3 at room temperature with $J \sim 7$, the RFR will occur in the range of positron energies $\varepsilon \sim 15$ meV. This energy is much lower than what can currently be measured with the positron beam. However, a resonance at such low energy will certainly affect $Z_{\rm eff}$ values for *thermalized* positrons. Indeed, using the same value $\varepsilon_b = 2$ meV, we obtain room-temperature Maxwellian averages of $Z_{\rm eff}^{\rm (dir)} = 450$ and $Z_{\rm eff}^{\rm (res)} = 480$. Here about 98% of $Z_{\rm eff}^{\rm (res)}$ is due to the RFR. This contribution significantly reduces the gap between the theoretical $Z_{\rm eff} = Z_{\rm eff}^{\rm (dir)} + Z_{\rm eff}^{\rm (res)} = 930$, and the experimental room-temperature value, $Z_{\rm eff} = 1300$ [21].

5. Inversion doubling, overtones and combinations

As seen from the estimates in Secs. 3 and 4, molecular rotations cannot account for the discrepancy between the theory for mode-based VFRs and experiment. It appears that there must be some additional resonances which boost the annihilation rate below 0.15 eV and around 0.3 eV.

The spectrum of low-lying vibrational levels of ammonia is well known, both experimentally and theoretically (see, e.g., Refs. [22, 23]). In particular, NH_3 is a classic example of a molecule with inversion doubling. This occurs as a result of tunneling of the nitrogen atom through the plane of the hydrogen atoms. Thus, each vibrational level is split into two sublevels, a lower positive (+) and an upper negative parity level (−) [17]. The selection rule for infrared absorption, $+ \leftrightarrow -$, means that in spite of the doubling of the modes, only four fundamental transitions will be allowed from either component of the vibrational ground state (split by 0.79 cm^{-1}). However, if positron capture is accompanied by $+ \leftrightarrow +$ or $- \leftrightarrow -$ transitions, the number of mode-based resonances accessible from either component of the ground state doubles.

In addition, the energy range of interest (0–4100 cm^{-1}) contains a few overtones and combination vibrations. To assess the effect of their VFRs on the annihilation rate, we assume that these resonances have elastic widths $\Gamma_\nu^e \gg \Gamma^a$. We also moderate the contribution of overtones and combinations (somewhat arbitrarily) by the factor $\Gamma_\nu^e/\Gamma_\nu = 1/n$, where n is the total number of vibrational quanta in the excitation. A possible reason for such weighting is that a multiquantum vibrational excitation allows the positron to escape through a number of (vibrationally inelastic) detachment channels. Using the data from Table 2 of Ref. [23] and

Table 3 of Ref. [22], a total of 17 vibrational excitations are included, namely: ν_2 (+ and −), $2\nu_2$ (+ and −), $3\nu_2$ (+ and −), $4\nu_2$ (+ and −), ν_4, $\nu_2+\nu_4$ (+ and −), $2\nu_4$ (A_1), $2\nu_4$ (E), ν_1, ν_3, and $2\nu_2+\nu_4$ (+ and −).

The result of this calculation (which also included rotational broadening, as described in Sec. 3) is shown in Fig. 1 by the dot-dashed curve. The additional VFRs have built up spectral weight around 0.1 eV, and around 0.3 eV, i.e., in the gap between the modes. The calculation appears to be in better overall agreement with experiment, except for the strong overestimation of the peak at 0.4 eV. This suggests that the vibrational excitations listed above do indeed play a role in VFR and contribute to the positron annihilation. It is also clear that in ammonia one needs to calculate, rather then estimate, the elastic and total widths. This must be done for both infrared-allowed and infrared-forbidden (but "positron-allowed") transitions, including the inelastic escape channels for overtones and combinations.

6. Conclusions

Positron annihilation in ammonia has been analyzed using the resonant annihilation theory. Comparison with experiment indicates that vibrational resonances due to inversion doubling, as well as overtones and combination vibrations, all contribute to the annihilation rate for positron energies below 0.5 eV.

Due to the large rotational constants, the vibrational resonances in ammonia are significantly broadened by molecular rotations at room temperature. The permanent dipole moment (or, equivalently, the ground-state inversion doublet) gives rise to a rotational Feshbach resonance at sub-thermal positron energies. This resonance plays an important role in enhancing annihilation for room-temperature positrons.

It is possible that rotational Feshbach resonances can also occur in other small polyatomic molecules with small positron binding energies. In nonpolar molecules, the low-energy positron can couple to molecular rotations by means of the electrostatic quadrupole interaction.

References

[1] Smith P M and Paul D A L 1970 *Can. J. Phys.* **48** 2984
[2] Surko C M, Passner A, Leventhal M and Wysocki F J 1988 *Phys. Rev. Lett.* **61** 1831
[3] Gribakin G F 2000 *Phys. Rev. A* **61** 022720
[4] Gribakin G F 2001 *New Directions in Antimatter Chemistry and Physics* (Kluwer Academic, Dordrecht) chap 22, p 413 (Editors C. M. Surko and F. A. Gianturco)
[5] Gilbert S J, Barnes L D, Sullivan J P and Surko C M 2002 *Phys. Rev. Lett.* **88** 043201
[6] Barnes L D, Gilbert S J and Surko C M 2003 *Phys. Rev. A* **67** 032706
[7] Barnes L D, Young J A and Surko C M 2006 *Phys. Rev. A* **74** 012706
[8] Young J A and Surko C M 2008 *Phys. Rev. A* **77** 052704
[9] Young J A and Surko C M 2008 *Phys. Rev. A* **78** 032702
[10] Danielson J R, Young J A and Surko C M 2009 *J. Phys. B* **42** 235203
[11] Gribakin G F and Lee C M 2006 *Phys. Rev. Lett.* **97** 193201
[12] Young J A, Gribakin G F, Lee C M R and Surko C M 2008 *Phys. Rev. A* **77** 060702
[13] Barnes L D 2004 *Vibrational Feshbach resonances in positron annihilation rates for molecules* Ph.D. thesis University of California, San Diego
[14] Bishop L M and Cheung L M 1982 *J. Phys. Chem. Ref. Data* **11** 119
[15] Gribakin G F 2002 *Nucl. Instrum. Methods Phys. Res. B* **192** 26
[16] Gribakin G F and Lee C M R 2006 *Nucl. Instrum. Methods Phys. Res. B* **247** 31
[17] Herzberg G 1945 *Molecular spectra and molecular structure: II. Infrared and Raman spectra of polyatomic molecules* (Van Nostrand, New York)
[18] Poynter R L and Kakar R K 1975 *Astrophys. J. Suppl. Ser.* **29** 87
[19] Lide D R (ed) 2000 *Handbook of Chemistry and Physics* 81st ed (CRC, Boca Raton)
[20] Garrett W R 1975 *Phys. Rev. A* **11** 509
[21] Heyland G R, Charlton M, Griffith T C and Wright G L 1982 *Can. J. Phys.* **60** 503
[22] Handy N C, Carter S and Colwell S M 1999 *Mol. Phys.* **96** 477
[23] Rashev S and Moule D C 2006 *J. Mol. Spectrosc.* **235** 93

Relaxed geometries and dipole moments of positron complexes with diatomic molecules.

Denise Assafrão and José R. Mohallem
Laboratório de Átomos e Moléculas Especiais, Departamento de Física, ICEx, Universidade Federal de Minas Gerais, CP 702, 30123-970, Belo Horizonte, MG, Brasil
E-mail: rachid@fisica.ufmg.br

Abstract. Relaxed geometries and dipole moments of diatomic molecules interacting with a slow positron are reported as functions of the positron distance to the more electronegative atom. A molecular model for the complex that allows applications to large systems is used. The electron population on the positron is proposed as a weighting function to calculate the average quantities. Results show Self-Consistent-Field quality or better.

1. Introduction

After some controversy [1, 2], it is now well accepted that a slow energy binding positron is able to alter the equilibrium geometries of molecules and, in consequence, their geometry-dependent properties like dipole moments. Some studies in this direction appeared in the recent literature, though restricted to diatomic molecules [3, 4] due to computational obstacles. However the possibility of extending this important study to larger molecules is now opened with the development of the molecular model [5]. In fact, we reported recent applications to positron complexes with the H_2O, CH_3F and CH_3Cl molecules [6]. In that work, however, we limited to present results for relaxed molecular properties as functions of the positron distance.

A relevant progress would be to find a feasible way of averaging the molecular properties over positron distances without having to calculate the (for now unfeasible) positron wavefunction. In the present work we propose the electron population on the positron as a possible weighting function to perform the averages. Application to geometry relaxation of some diatomic molecules, namely XH (X=Li, Na, K) and BeO, display Self-Consistent-Field (SCF) accuracy or better. The relaxed molecular dipole moments are then calculated as well, in view of their relevance to the study of positron interaction with molecules.

2. Procedure

We perform nuclear-mass-dependent-type Born-Oppenheimer electronic calculations already fully described [5, 6]; here we present just the main features of the approach. Electronic orbitals are expanded in basis functions centered on the nuclei and also on the positron, which is taken here as a light nucleus. A positron-mass-dependent kinetic energy term like

$$-\sum_{i}^{n} \frac{\nabla_i^2}{2m_p} \qquad (1)$$

is added to the matrix elements of the n-electron Born-Oppenheimer Hamiltonian, whenever these matrix elements involve two basis functions centered on the positron. In Eq. (1) m_p is the positron mass and the index i refers to the electrons. Matrix elements of crossing orbitals are kept unchanged. This procedure guarantees that the kinetic energy of all electrons is corrected for the reduced mass effect when they occupy an orbital centered on the positron. Identical terms depending on the nuclear masses can be added as well but they become irrelevant here, in view of the smallness of the positron mass which yields a huge positron correction.

We are then able to map the electronic energy for different configurations of the nuclei-positron systems by following the standard Born-Oppenheimer procedure. Adding to it the repulsion energy of the positive charged particles we thus generate potential energy surfaces (PES) for positron and nuclei motions. For each positron position the molecule is allowed to relax its geometry and, in consequence, the geometry dependent properties. Other typical electronic quantities as populations, dipole moments, etc., are available from the corresponding computational code [7].

The above procedure is equivalent to an adiabatic separation of the positron and electrons motions. But since we keep the positron kinetic energy the only strong condition to be satisfied by this approach is that the PES in question, namely the ground state PES, is isolated from the others (no crossings or pseudo-crossings involved). We noted that this condition hold for all systems we studied so far.

3. Results and discussion

Figure 1 shows schematically the framework and relevant coordinates we use. We observed that the positron occupies a kind of lobe orbital centered on the more electronegative atom, H or O. Up to $\theta = 60$ degrees, with a fixed R_p, the PES energy is practically insensitive to θ for all systems, but starts rising substantially for θ approaching 90 degrees. In consequence we found appropriate to neglect the contribution of configurations having $\theta > 90$ degrees and also assume axial symmetry around the z axis. For the alkaline compounds the typical behavior of the electronic density (represented in figures 2 and 3 for NaH-e^+) shows the transference of almost the whole molecular bond to bind e^+ to H, while for BeO-e^+ the electronic density shows no significant change. The global minima of the PES for all systems lie on this axis so that we plot cuts along z of the PES, the interatomic distances and the Löwdin electronic population [8] on the positron against positron distance, shown in figure 4.

A common behavior for the alkaline compounds is that the interatomic distance increases as the positron approaches and reaches its maximum close to the minimum of the PES, while for BeO-e^+ the interatomic distance is again practically insensitive to the positron. This feature allows large relaxation of interatomic distances and properties for the alkaline compounds.

Due to the large mobility of the positron, the PES minima are hardly useful to represent relaxation. In order to do it without calculating the positron wavefunction we need a kind of distribution function to weight geometries and properties in the generation of reliable averages. In this work we consider the Löwdin electronic population $\eta(R_p)$ on the positron as a candidate to be this function. This quantity, defined as

$$\eta(R_p) = \sum_{\nu} P_{\nu\nu}, \tag{2}$$

where **P** is the density matrix for any MO-LCAO (Molecular Orbital-Linear Combination of Atomic Orbitals) [9] is choosen in view of its better behavior at shorter positron-molecule distance compared to the more traditional Mülliken population [9], as observed in this work.

Any molecular property M is then evaluated as

$$< M >= \frac{\int \eta(R_p) M(R_p) R^2 dR_p}{\int \eta(R_p) M(R_p)}. \tag{3}$$

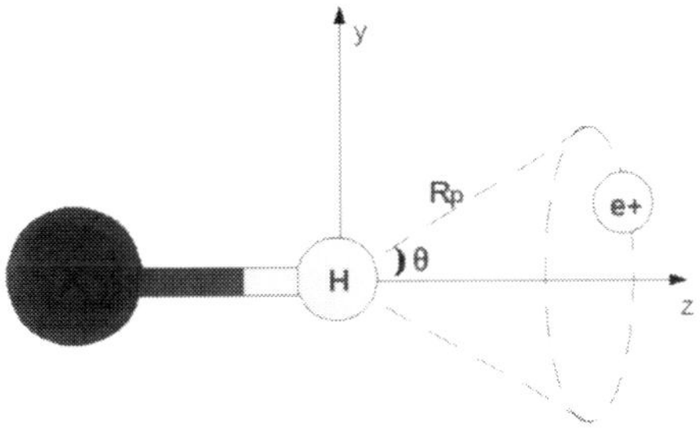

Figure 1. Schematic representation of the XH-e$^+$ (X = Li, Na e K) complexes. The minima of the PES lie on the z-axis. A similar framework applied to BeO-e$^+$. R$_p$ is the distance of the positron to the H atom and θ is the axial angle.

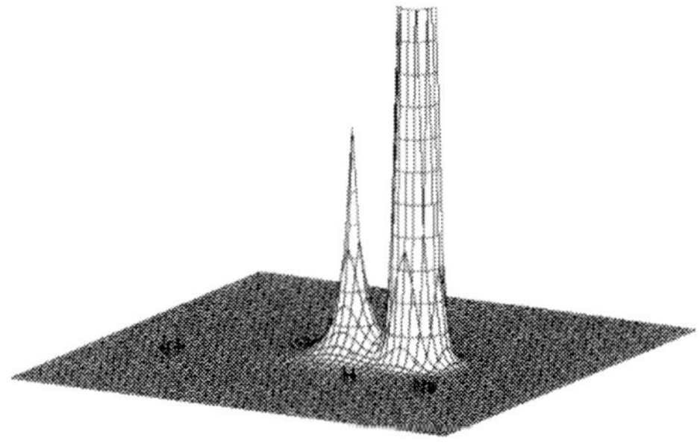

Figure 2. Charge density of the NaH-e$^+$ complex for R$_p$=10a_o.

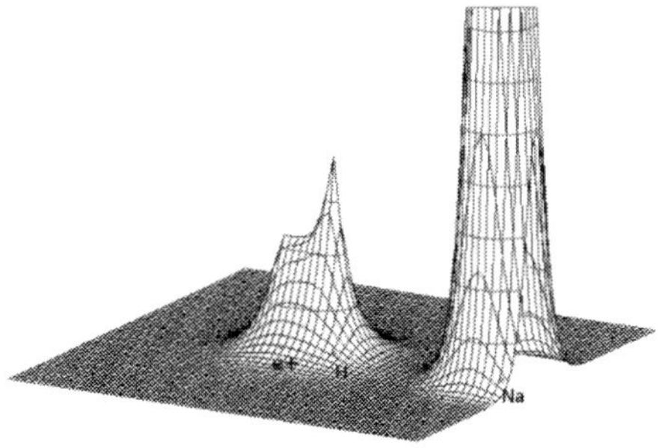

Figure 3. Charge density of the NaH-e$^+$ complex in the minimum of the PES.

The integrals in Eq. (3) are evaluated numerically on a grid truncated at R$_p$ = 10 a_o (already at 5 a_o we verified that the electronic density on the positron becomes negligible).

The electronic populations in figure 4 are considered as weighting functions and the relaxed dipole moment is evaluated for the isolated molecules in the relaxed geometries. This means that for each positron position we obtain the new minimum molecular geometry; in these relaxed geometry we calculate the dipole moment of the isolated molecule. Except for BeO-e^+ we observe significant molecular relaxation as the positron approaches. Particularly the dipole moments increase significantly as the positron approaches the minima of the PES. Using Eq. (3) we average internuclear distances and dipole moments, and compare the first ones with sophisticated values obtained in full correlated calculations [3, 10, 11], displayed in Table 1.

As expected, when we just locate the positron on the minimum of the PES the relaxed geometries look too large so that the corresponding dipole moments (both carrying the subscript "min" in the table) are not reliable. Account must be taken to the positron delocalization. This is accomplished first by the calculation of the average values with Eq. (3). We note that for LiH-e^+ our result approaches the SCF one, recovering 40% of the relaxation reported in the assumed best calculation of Strasburger [10]. But for the two other molecules our results lie between the

Journal of Physics: Conference Series **199** (2010) 012014 doi:10.1088/1742-6596/199/1/012014

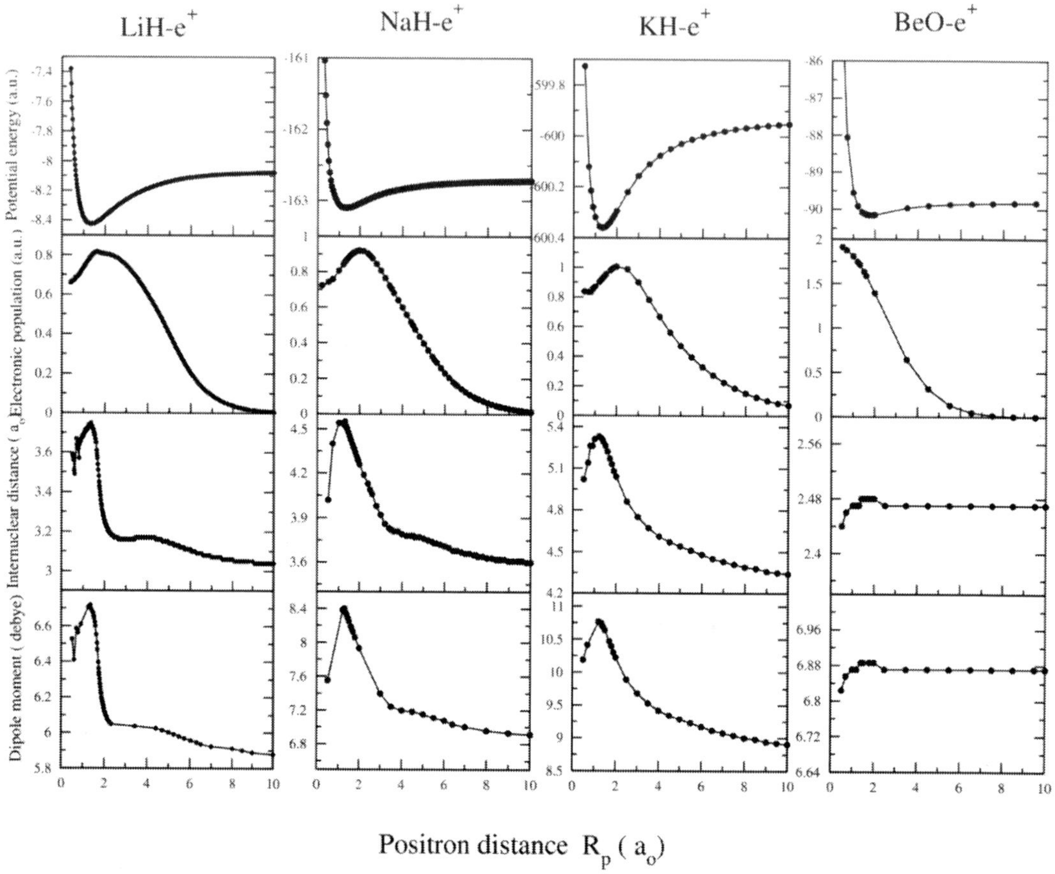

Figure 4. Interatomic distances (XH and BeO), potential energy and electronic population on the positron of the XH-e$^+$ (X = Li, Na e K) and BeO-e$^+$ complexes against positron distance.

SCF and the Configuration Interaction (CI) results from Buenker et al [3]. The corresponding dipole moments increase by about 3-6% relative to the isolated ones as we go from the lighter to the heavier systems. As we did in [6] we evaluate the two quantities, for comparision, also at $Z_p = Z_{p,min} + \frac{1}{2}a_o^{Ps}$, a typical distance for positronic systems, where $a_o^{Ps} = 2a_o$ means the positronium Bohr radius. These quantities are represented in the table by the subscript "ave" and show an improvement relative to the previous ones as we consider the CI results from [3]. Again the dipole moments show an enhancement relative to the unperturbed ones, now of 3-15%. It becomes clear that the study of the binding of a positron to these molecules should take relaxation in account.

As for the BeO-e$^+$ complex we found no noticeable relaxation, in full agreement with the literature [4], suggesting that the positron binding to this molecule should be explained by its unperturbed dipole moment as well as by the positron affinity of the oxygen atom along.

Table 1. Internuclear distance ($R_{min}/\ a_o$) and dipole moment (μ_{min}/Debye) in the minimum of the PES, electronic population averaged internuclear distance ($\langle R\rangle/a_o$) and dipole moment ($<\mu>$/Debye) and internuclear distance (R_{ave}/a_o) and dipole moment (μ_{ave}/Debye) in the positron averaged position. [a] Reference [3]. [b] Reference [10]. [c] Reference [11].

	R_{min}	μ_{min}		R_{min}	μ_{min}	$\langle R\rangle$	$<\mu>$	R_{ave}	μ_{ave}
LiH	**3.012**	**5.858**	LiH-e^+	**3.75**	**6.718**	**3.152**	**6.004**	**3.18**	**6.043**
						3.209(SCF)[a]			
	3.045[a]	5.90[a]				3.391(CI)[a]			
						3.348[b]			
	3.015[c]					3.458[c]			
NaH	**3.550**	**6.839**	NaH-e^+	**4.52**	**8.352**	**3.795**	**7.182**	**4.15**	**7.751**
						3.63(SCF)[a]			
	3.597[a]	6.901[a]				4.115(CI)[a]			
KH	**4.231**	**8.693**	KH-e^+	**5.31**	**10.728**	**4.533**	**9.246**	**4.90**	**9.977**
						4.365(SCF)[a]			
	4.250[a]	8.618[a]				5.063(CI)[a]			

Some speculative considerations on the positron bonding mechanism to polar diatomic molecules ensue from our calculations. The calculated binding energies of the four molecules considered here have magnitudes of 10^{-1} eV [3, 10, 11], about two orders of magnitude higher than slow projectile positrons prepared in experiments [12]. So, it seems reasonable that a slow positron will be easily trapped in the molecular potential while its produces a torque on a non-rotating molecule in order to align with the molecular axis in its preferential site, bonded to the more electronegative atom. Both during the bonding process as well as after binding takes place, the molecular relaxation of geometry and properties must be taken in account, as we consider its large incursions allowed by its huge kinetic energy (as compared to the other nuclei). As a matter of fact, the relaxation effects studied here have a static character. This means that once one starts considering the dynamics of the positron-molecule interaction, the molecular vibrations must be taken from the relaxed minimum geometries, instead of the geometries of the isolated molecules; the exception here is BeO-e^+. These conclusions affect not only the binding energy considerations but also other features like annihilation rates and low energy cross sections (cases in which the scattering can be described by the PES).

In conclusion, the molecular approach can handle well the problem of molecular relaxation under positron interaction without the computational burden that obstacles the application of full correlated methods. In the applications we obtained a systematic enhancement of the molecular dipole moments of diatomic hydrides subject to positron binding. The electronic population on the positron is proposed as a weighting function to evaluate average molecular quantities over positron positions. On the other hand, the best results obtained with the typical distance $Z_p = Z_{p,min} + \frac{1}{2}a_o^{Ps}$ suggest that better weighting functions could still be attempted. We will keep investigating this point.

3.1. Acknowledgments

We would like to acknowledge the brazilian agencies Conselho Nacional de Pesquisas (CNPq) and FAPEMIG for financial support to this work.

Journal of Physics: Conference Series **199** (2010) 012014 doi:10.1088/1742-6596/199/1/012014

4. References

[1] Mohallem J R and Gonçalves C P, J. Chem. Phys. **121**, 5553 (2004).
[2] Tachikawa M, Buenker R J and Kimura M, J. Chem. Phys. **121**, 9191 (2004).
[3] Buenker R J, Liebermann H, Melnikov, Tachikawa M, Pichl L and Kimura M, J. Phys. Chem. A **109**, 5956 (2005).
[4] Gianturco F A, Franz J, Buenker R J, Liebermann H, Pichl L, Rost J, Tachikawa M and Kimura M, Phys. Rev. A **73**, 022705 (2006).
[5] Mohallem J R, Rolim F and Gonçalves C P, J. Phys. B. **37**, 1045 (2004).
[6] Assafrão D and Mohallem J R, submited to Phys. Rev. A.
[7] Gonçalves C P and Mohallem J R, J. Comput. Chem. **25**, 1736 (2004).
[8] Löwdin P O, J. Chem. Phys. **18**, 365 (1950).
[9] Szabo A and Ostlund N S, *Modern quantum chemistry: introduction to advanced electronic structure theory*, Dover Publications, Inc, Mineola, New York (1996).
[10] Strasburger K, J. Chem. Phys. **114**, 615 (2001).
[11] Mella M, Morosi G, Bressanini D and Elli S, J. Chem. Phys. **113**,6154 (2000).
[12] See, for example, J. A. Young e C. M. Surko, Phys. Rev. A **78**, 032702 (2008).

Elastic scattering of slow electrons by positronium atoms in weakly-coupled plasmas

Arijit Ghoshal[1,*], Sabyasachi Kar [2], and Y. K. Ho[1]
[1]Institute of Atomic and Molecular Sciences, Academia Sinica, Taipei, Taiwan 106, R.O.C
[2]Center for Theoretical Atomic and Molecular Physics, The Academy of Fundamental and Interdisciplinary Sciences, Harbin Institute of Technology, Harbin 150080, China
*Permanent address: Department of Mathematics, Burdwan University, Golapbag, Burdwan 713 104, West Bengal, India.

E-mail: ykho@pub.iams.sinica.edu.tw

Abstract
Low-energy elastic scattering of electrons from the ground state of positronium atoms embedded in weakly-coupled plasmas has been studied within the framework of an effective range theory. Plasma screening effect has been taken care of by screened Coulomb potential. Highly correlated and variationally determined wave functions for Ps^- are used to determine the effective range of the ion states. The results for *S*-wave singlet phase shifts for the incident electron momentum in the range of [0, 0.5] a.u. are reported, to the best of our knowledge, for the first time in the literature.

1. Introduction

Atomic collision processes in weakly-coupled plasmas (WCP) are currently interesting topic of research [1-3] (and further references therein). The potential of a test charge in a classical weakly-coupled (collision less) plasma is represented by Debye-Huckel model or screened Coulomb potential (SCP) [1]:

$$V(r) = (1/r)e^{-r/D} \quad \text{(in a.u.)}, \qquad (1)$$

where $D(=v_T/\omega_p)$ is the Debye length, v_T and ω_p are the thermal velocity and plasma frequency respectively. This is a good approximation for hot-dense plasmas and low-density warm plasmas. In this work, we focus our attention to the effect of screening of weakly-coupled plasmas on the low energy elastic scattering of electrons from the ground state of positronium atoms (Ps) within the framework of an effective range theory. In vacuum, among others, the low energy elastic scattering of electrons by the positronium atoms was carefully investigated by Ward *et al* [4] and Igarashi *et al* [5]. But in a plasma environment, to the best of our knowledge, the elastic phase shifts for electron-positronium scattering have not been reported so far. Recently Kar and Ho [6] have determined sufficiently accurate wave functions of positronium negative ion (Ps^-) in WCP using Ritz's variational principle. Effective range theory gives us an opportunity to calculate some scattering parameters at low incident energies from the knowledge of the bound state wave functions of the composite scattering system (Ps^-). The atomic units (a.u.) are used throughout the present work.

2. Theory

The phase shift δ in the singlet electron-positronium scattering is related to the momentum of the incident electron k_i by the relation [7]:

$$k_i \cot\delta = -\gamma + \frac{1}{2} r_0(\gamma^2 + k_i^2) + O(k_i^4) \tag{2}$$

where γ is related to the binding energies of Ps^- and Ps, $\varepsilon_{Ps^-}, \varepsilon_{Ps}$, by

$$\gamma = \sqrt{4(\varepsilon_{Ps} - \varepsilon_{Ps^-})/3} \tag{3}$$

and r_0 is a constant of the dimension of a length, which is called the "effective range" of the ion state. Let $\Psi(\mathbf{r}_1, \mathbf{r}_2, \mathbf{r}_{12})$ denotes the ground state wave function of Ps^-, where $\mathbf{r}_1, \mathbf{r}_2$ are the position vectors of the two electrons relative to the positron and $\mathbf{r}_{12}$ is their relative coordinate. Then representing Ψ in the asymptotic form:

$$\Psi(\mathrm{r}_1 = 0, |\mathrm{r}_2| = r) = \Psi(|\mathrm{r}_1| = r, \mathrm{r}_2 = 0) = \Psi(r) \to C\frac{e^{-\gamma r}}{r}, \tag{4}$$

it can be shown that r_0 is given by [7,8]

$$r_0 = \frac{1}{\gamma} - \frac{1}{4\pi C^2}. \tag{5}$$

The constant C can be calculated from the formula [8]

$$C = Gre^{\gamma r}\Psi(r, 0, r), \tag{6}$$

where G is a normalization constant. Then it is seen that over a wide range in the asymptotic region $C(r)$ remains constant. Having obtained the constant value of C, we obtain the effective range r_0 from relation (5). This r_0, when substituted in relation (2), gives the scattering length and phase shift. In order to determine the ground state energies and wave functions of Ps^- in WCP, we use Ritz's variational principle and employ the following correlated wave function [6]:

$$\Psi = \frac{1}{\sqrt{2}}(1 + P_{12})\sum_{i=1}^{N} A_i \exp(-\alpha_i r_1 - \beta_i r_2 - \delta_i r_{12}), \tag{7}$$

where $\alpha_i, \beta_i, \delta_i$ are non-linear variational parameters, $A_i (i = 1, ..., N)$ are the linear expansion coefficients and P_{12} is the permutation of the two electrons. The ground state energies and wave functions of Ps atom in WCP are also determined within the framework of Ritz's variational principle by employing the wave function of the form:

$$\Psi = \sum_i A_i e^{-\alpha_i r_{12}} r_{12}^{l_i}, \quad l_i = 0, 1, 2, ... \tag{8}$$

where A_i's are linear expansion coefficients and α_i's are non-linear variational parameters.

3. Results and discussions

Using the ground state energies of Ps^- and Ps we first calculate γ which is then used to compute the constant C. In Figure 1(a) we plot C as a function of r, for the Debye length 50, for three different expansions of the wave function (7). From these figures it is seen that a stable plateau for C exits in the r coordinates ranging from 20 to 45 a. u.. In Figure 2(b) we have shown the constant C for various values of the screening parameter $\mu (= 1/D)$. From this figure it is seen that C decreases steadily with increasing screening effect. Having determined γ and C, we calculate the scattering lengths and phase shifts by using formula (2). In Table 1 we compare our present result for the unscreened case with the variational result of Ward *et al* [4] and close-coupling result of Igarashi *et al* [5]. The comparison establishes the accuracy of our results for the phase shifts. In the present work, for each D we retain our calculation for the incident electron momentum in the range for which only the elastic channel is open, i.e., below the first excitation threshold of Ps atom. So that effective range theory is expected to yield reasonably accurate results. The overall uncertainties in our reported results are estimated to within 10% for the highest energy reported here.

Table 1: Singlet (S=0) phase shifts (in radian) for the elastic scattering of electrons by Ps(1s) in vacuum.

	*k_i=0.0	k_i=0.05	k_i=0.1	k_i=0.15	k_i=0.2	k_i=0.25	k_i=0.3	k_i=0.4	k_i=0.5
Present	12.242	2.552	2.063	1.695	1.420	1.213	1.054	0.829	0.680
(A)	12.38(7)	2.547(4)	2.049(1)	1.668(2)	1.378(2)	1.156(2)	0.984(2)	0.748(2)	0.617(7)
(B)			2.056		1.378		0.983	0.746	

* Scattering length. (A): Result of Ward *et al* [4]; Result of Igarashi *et al* [5].

Table 2: Singlet (S=0) phase shifts (in radian) for the elastic $e^- - Ps(1s)$ scattering in WCP.

	k_i (a.u.)									
D	0.0*	0.05	0.1	0.15	0.2	0.25	0.3	0.35	0.4	0.5
∞	12.242	2.552	2.063	1.695	1.420	1.213	1.054	0.929	0.829	0.680
50	12.265	2.551	2.062	1.694	1.420	1.206	1.054	0.929	0.829	0.680
15	12.493	2.541	2.050	1.682	1.410	1.204	1.048	0.924	0.825	0.677
14	12.532	2.540	2.048	1.680	1.409	1.196	1.047	0.923	0.824	0.676
10	12.783	2.529	2.034	1.668	1.399	1.193	1.040	0.918	0.819	0.672
9	12.896	2.524	2.028	1.663	1.399	1.188	1.038	0.915	0.817	0.671
8	13.054	2.518	2.020	1.656	1.389	1.183	1.034	0.912	0.815	0.669
7	13.273	2.509	2.010	1.647	1.382	1.176	1.030	0.909	0.812	0.667
6	13.594	2.496	1.995	1.635	1.373	1.171	1.025	0.906	0.810	0.666
5	14.101	2.477	1.974	1.620	1.363	1.177	1.022	0.904	0.809	0.667
4	14.974	2.446	1.947	1.606	1.362	1.177	1.033	0.918	0.824	
3	16.689	2.401	1.942	1.656	1.452	1.294	1.166	1.059	0.968	
2.4	18.673	2.401	2.098	1.974	1.916	1.889	1.878	1.876		
2	20.495	2.607	2.661	2.757	2.830	2.882	2.920			
1.9	20.646	2.769	2.875	2.948	2.991	3.019	3.039			
1.8	20.718	2.968	3.042	3.073	3.090	3.100	3.107			

* Scattering length.

In Table 2 we list the present S-wave singlet phase shifts for elastic $e^- - Ps(1s)$ collision in WCP for various Debye lengths. From this table we see that the scattering length increases with increasing screening effect. This is quite expected. Because, the effective scattering area (area of the collision region) placed by the target atom increases with increasing screening effect. As a consequence the total scattering cross section increases. But at the zero energy, scattering predominantly occurs at S-wave, and cross section is proportional to the square of the scattering length. Consequently the scattering length increases with increasing screening effect.

4. Conclusions

In the present work, we have investigated the plasma screening effect of weakly-coupled plasmas on the low energy electron-positronium elastic scattering using an effective range theory. Effective range theory is very effective for low energy $e^- - Ps(1s)$ scattering, because the electron affinity of Ps is very small compared with the kinetic energy of the electron inside the Bohr radius (validity of effective range theory in the low energy region [7]). We have found that the scattering length increases with decreasing Debye length. We report the results for phase shifts for various screening effects for the incident electron momentum in the range for which only elastic channel is open. For the pure Coulomb case, our results compare well with the best results in the literature, with estimated uncertainly within 10% for the highest energy reported in the present work. The results for the screened Coulomb case are reported, to our knowledge, for the first time in the literature. We hope that our results will provide useful information to the research communities in plasma physics, astrophysics and positron physics.

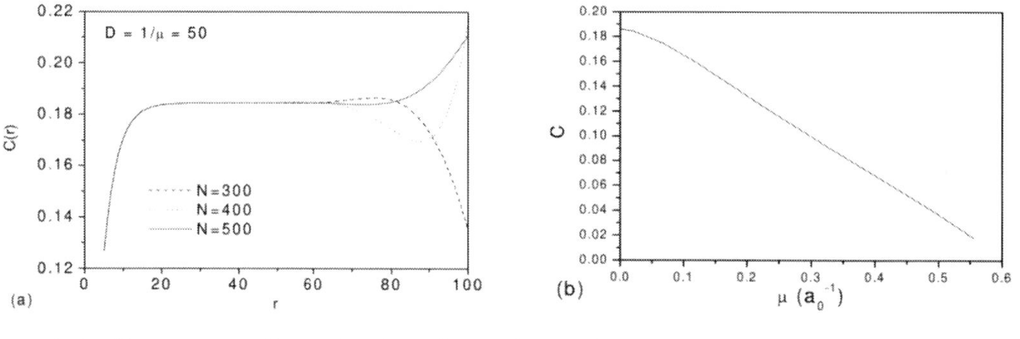

Figure 1. (a) The constant C as a function of coordinate r for different expansion lengths of the wave function (7) D = 50. (b) The constant C as a function of the screening parameter μ in WCP.

References

[1] Weisheit J C 1988 *Adv. At. Mol. Phys.* **25** 101.
[2] Murillo M S 1996 *Atomic Processes in Plasmas, AIP Conf. Proc. No. 381*, edited by A. L. Osterheld and W. H. Goldstein, (American Institute of Physics, Woodbury, NY) p. 231.
[3] Ghoshal A and Ho Y K 2009 *Phys. Rev. A* **79** 062514; *J. Phys. B: At. Mol. Opt. Phys.* **42** 075002.
[4] Ward S J, Humberston J W and McDowell M R C 1987 *J. Phys. B: At. Mol. Opt. Phys.* **20** 127.
[5] Igarashi A, Nakazaki S and Ohsaki A 2000 *Phys. Rev. A* **61** 032710.
[6] Kar S and Ho Y K 2005 *Phys. Rev. A* **71** 052503; 2006 *Phys. Rev. A* **73** 032502.
[7] Ohmura T, Hara Y and Yamanouchi T 1958 *Progr. Theo. Phys.* **20** 82.
[8] Bhatia A K and Drachman R J 1985 *Phys. Rev. A* **32** 3745.

On new developments in the physics of positron swarms

Z. Lj. Petrović[1], A. Banković[1], S Dujko[1], S. Marjanović[1], M. Šuvakov[1], G. Malović[1], J. P. Marler[2], S.J Buckman[3], R.D. White[4] and R.E. Robson[4]

[1] Institute of Physics Belgrade, University of Belgrade, Pregrevica 118, POB 68, 11080 Zemun Serbia

[2] University of Aarhus, Aarhus Denmark

[3] ARC Centre for Antimatter-Matter Studies, Research School of Physical Sciences and Engineering, The Australian National University, Canberra, ACT, Australia

[4] ARC Centre for Antimatter-Matter Studies, School of Mathematics, Physics and IT, James Cook University, Townsville 4810, QLD, Australia

zoran@phy.bg.ac.rs

Abstract. Recently a new wave of swarm studies of positrons was initiated based on more complete scattering cross section sets. Initially some interesting and new physics was discovered, most importantly negative differential conductivity (NDC) that occurs only for the bulk drift velocity while it does not exist for the flux property. However the ultimate goal was to develop tools to model positron transport in realistic applications and the work that is progressing along these lines is reviewed here. It includes studies of positron transport in molecular gases, thermalization in generic swarm situations and in realistic gas filled traps and transport of positrons in crossed electric and magnetic fields. Finally we have extended the same technique of simulation (Monte Carlo) to studies of thermalization of positronium molecule. In addition, recently published first steps towards including effects of dense media on positron transport are summarized here.

1. Introduction

Over the past 3 years the field of positron swarms has been revitalized. The basis for the new series of papers was the recently measured cross sections for binary collisions [1-5] supported by equally sophisticated calculations [6,7]. The early studies of positron swarms produced a limited range of data due to technical difficulties in performing experiments and limited apparatus developed to treat non-conservative transport [8,9]. Thus we have only drift velocities for several gases [10,11] H_2, CO_2 and O_2. Drift velocities (W) for hydrogen were obtained at elevated *E/N* and show a surprising display of negative differential conductivity (NDC) which was explained [8] as being the result of mean energy dependent effective Z as the observable in the experiment is *W/Z*. It was uncertain whether other gases displayed NDC since the range of *E/N* covered did not extend sufficiently towards higher mean energies. It must be noted that theoretical-numerical calculations of transport coefficients for positrons were not performed at that stage as data for the cross sections were lacking and the experimental database was narrow and perhaps did not carry over to rare gases which would be easier to study. However, related to swarm studies and perhaps even belonging to swarm studies were the calculations and measurements of thermalization times and lifetimes of positrons in gases [12-14] which were performed continuously in spite of the demise of the other swarm studies.

Associated with swarm studies are the developments of gas filled positron traps [9,15] where collisions are used to thermalize positrons below the onset of Ps formation and even further down to energies below 100 meV. While those systems may not be viewed as primarily 'swarm' like systems, the basic description of the collisions and how a thermalized ensemble is formed can be described by swarm physics and phenomenology. In this paper we show results obtained by a Monte Carlo simulation with complete sets of cross sections for the low energy (below 20 eV) range as described in [16-18].

2. Recent results on swarms of positrons

The new series of papers on transport of positrons in gases has been primarily focused on revealing the transport properties under the influence of excessive non-conservative (number changing) processes. Positronium formation is at least two orders of magnitude larger than electron attachment and it has a sharp rise after the threshold. The most important result was the observation of very large negative differential conductivity (NDC) which occurs only for the bulk drift velocity.

There are two kinds of transport coefficients: the first, flux, which enter through the flux-gradient relation and the second, bulk, which are to be used in the diffusion equation. The difference between the two sets arises only due to energy-dependent number-changing processes such as attachment and ionization for electrons and Ps formation for positrons. The relationship between the flux and bulk drift velocities may be calculated from the rate of the non-conservative process (in this case v_{PF} the Ps formation rate) [15]:

$$W_B = W_F - \frac{2\varepsilon}{3e}\frac{dv_{PF}}{dE}$$

In case of the NDC that was found for positrons [16,17], the special and unusual features were that (i) no NDC was present or even likely for the flux property, (ii) the difference between flux and bulk properties were as large as two orders of magnitude and (iii) it did not involve energy controlling processes, such as inelastic cross sections. The effect on the drift velocity and a similar effect for diffusion are caused only by the nature of the Ps formation. If there are inelastic processes in competition with Ps formation, such as in the case of nitrogen, the NDC practically disappears [18]. Nothing similar to this has been found for electrons but a general prediction of the effect was made [19] and it was found that for electrons this effect was not likely to occur.

In this paper we give a brief review of the more recent developments in the physics of positron swarms.

3. Positrons in liquids

The most interesting new development is the theory for the representation of charged particle transport in liquids [20]. This involved adjusting the magnitude and the shape of the cross section to account for coherent scattering off correlated particles. While this approach is valid only for non-polar molecules and includes only one of several possible effects occurring at high pressures this is a major first step and it revealed some of the important aspects of the physics that is involved. For example, as expected by the dependence of the de Broglie wavelength on particle energy the effect is pronounced at low energies, well below the threshold for Ps formation. The momentum transfer cross sections at energies up to a few eV are reduced considerably, leading to higher drift velocity in the liquid phase as compared with the gas phase [20] as well as other interesting effects for flux diffusion coefficients as shown in Figure 1. For drift velocity another form of NDC is induced purely by the structure of the medium, and its onset is well below the onset for the NDC caused by Ps formation and its non-conservative nature.

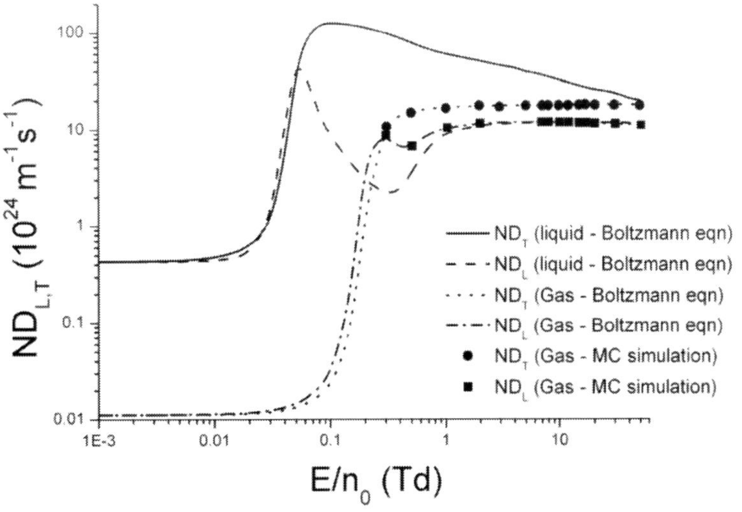

Figure 1. Diffusion coefficients for positrons in liquid argon calculated on the basis of cross sections given in [20] for liquid and [16] for gas phase.

It is essential to include other possible effects that may occur at high pressures and also to extend the theory to polar molecules, and these are the subject of current investigation. However, it is clear that at higher energies positrons move through a liquid in the same way as in a rarefied gas.

4. Positrons in crossed electric and magnetic fields

Most applications of positrons involve presence of high magnetic fields. In some cases the ability to tune independently the electric and magnetic fields may provide the desired degree of control over the properties of the ensemble. In any case even the first studies of transport (still unpublished but preliminary results were presented in [17]) have shown very interesting physical phenomena unlike any observed for electrons. The NDC in argon for example means that the bulk drift velocity (the drift velocity that includes the term due to non-conservative processes, is measured in actual experiments and is to be used in diffusion equation) may become very small, even two orders of magnitude smaller than the flux property (the property that is obtained by averages over velocity distribution and is to be used in flux gradient relation). At the same time there is an additional component of drift velocity when magnetic fields are present in the *ExB* direction. Unlike the *E* component, the *ExB* component of the drift velocity does not show a difference between flux and bulk properties (as there is no spatial segregation of particles along the *ExB* direction according to the energy). In the absence of NDC the *ExB* component is smaller than the *E* component. In the presence of NDC however, W_E may become very small and the overall W is dominated by W_{ExB}. Thus the total drift velocity W may not show NDC. Nevertheless it still has a strong mark of the non-conservative nature of Ps formation. The bulk drift velocity will be practically at right angles to the electric field (equal to the *ExB* component) while the flux property in the absence of NDC will be dominated by the *E* component. Thus the deflection angle between effective bulk and flux drift velocities will be quite large and instead of NDC the bulk drift velocity in magnetic fields has an additional deflection. The effect is shown for magnetic deflection in hydrogen in Figure 2. At very high magnetic fields normalized by the gas number density (in units of Huxley; 1 Hx=10^{-27} Tm^3) the effect is reduced since deflection for the flux property is close to the right angle anyway so the non- conservative effect cannot change the direction very much [22]. The effect is also reduced at any effective magnetic field for nitrogen, which does not have the NDC [18, 21].

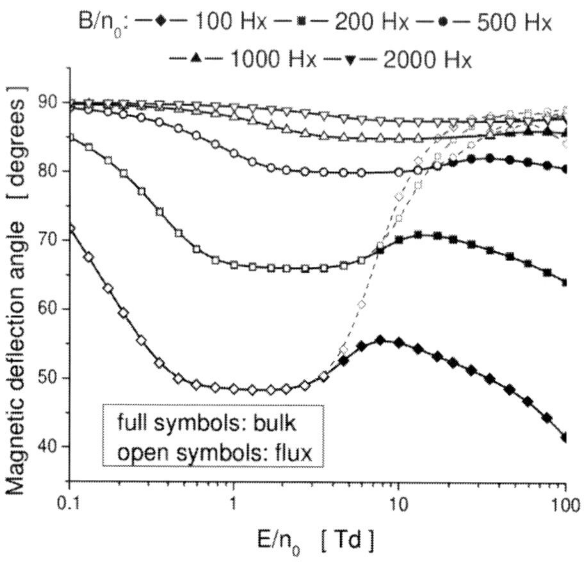

Figure 2. Variation of the bulk and flux values of the magnetic deflection angle for positrons in H_2 as a function of B/n_0 and E/n_0 in the crossed field configuration [21].

5. Thermalization of positrons and positronium in gases and in gas filled traps

One of the goals of the whole program is to model how positrons are thermalized and confined in gas filled traps. For an unbounded gas applying the primary sets of cross sections yields excellent agreement with experimental data for thermalization times [22].

Simulations for realistic geometry and other properties of a two stage Surko trap have been performed by the same technique. The conditions are: Stage I - N_2 p=10^{-3} torr L=100 cm; Stage II – (90%N_2+10%CF_4) p=10^{-4} torr L=100 cm; Barrier - 2eV; B=100 gauss; r=10cm – no losses on walls T_{gas}=300K ; 2000 initial particles; initial beam energy 10eV directed (+/- 1eV transverse). The results shown in Figure 3a present the decay of the mean energy and in Figure 3b the decay of the number of particles. A 2 eV barrier in the backward direction is imposed between the two stages. The mean energy in the first stage is reduced to energies of the order of 2-3 eV. This, relatively high value of mean energies in the first stage, after 10 micro seconds, is maintained by the high energy particles returning from the second stage through the 2 eV barrier. In the second stage, thanks to efficient vibrational energy losses, mean energies are reduced down to the thermal value (Figure 3a). The number of particles (Figure 3b) drops down very rapidly and the local peaks represent transfers of the ensemble between the two stages. Overall the number is reduced to 50% of the initial number of positrons. One should be warned, however, that we have not included annihilation in our set of data. This is justified for calculation of transport coefficients as the number of such collisions is very small and in the MC procedure may not occur at all or so rarely that their statistics is very poor. On the other hand when energies fall below the threshold for Ps formation, annihilation is the only loss process and over extended time periods (such as times of the order of 1 s as covered here) annihilation may

Journal of Physics: Conference Series **199** (2010) 012016 doi:10.1088/1742-6596/199/1/012016

contribute as much as Ps formation to the losses and needs to be included for realistic calculations of the efficiency of trapping [23]. This technique enables us to modify the procedure applied to the trap in order to optimize its performance, for example by adding radial rotating electric fields [24].

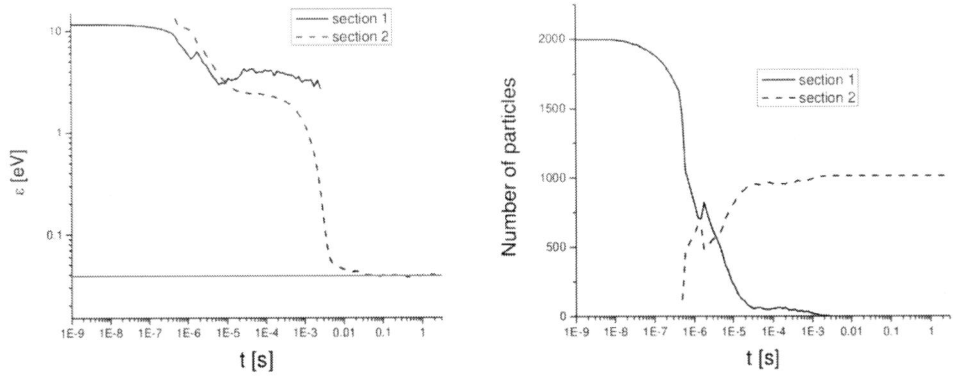

Figure 3. Temporal variation of the mean energy (a) and number (b) of positrons in a two stage Surko trap (conditions given in the text) [23].

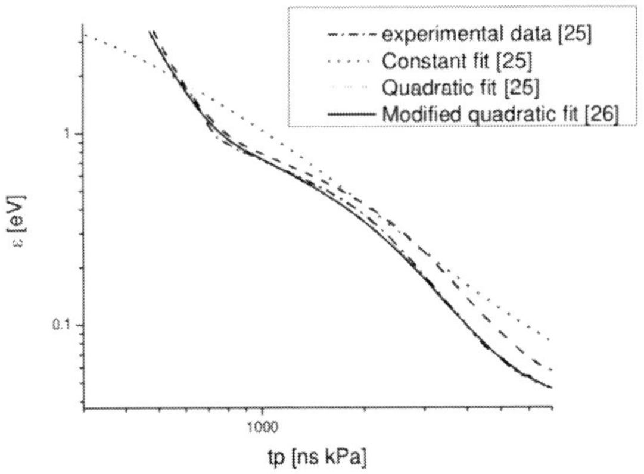

Figure 4. Thermalization of positronium (Ps) for different cross sections: dash-dot curve –experiment [25], solid black- modified quadratic fit [26], dotted and dashed- approximate theory [25].

Finally we mention the preliminary results on thermalization of Ps in He and water vapour. The experimental data were obtained in [25] and cross sections were derived based on approximate theory that assumes Maxwell Boltzmann distribution functions during the development of the Ps swarm. We have, on the other hand performed MC simulations and found that the distribution functions depart from a Maxwell Boltzmann distribution very rapidly due to selective removal of Ps in energy dependent cross sections [26]. Nevertheless the predicted thermalization (Figure 4) is still close to the experimental data requiring only small modifications of the cross section. Coupling of the codes for positrons and Ps would enable us to obtain the spatial profiles of the emitted gamma rays, which would be useful for the simulation of positron emission tomography (PET).

Acknowledgements
The work presented here was supported by MNTRS grant 141025. Some of the authors (SJB, RDW and RER) are also grateful to the Australian Research Council for support under the Centers of Excellence Program and to the Australian Government Department of Innovation, Industry, Science

Journal of Physics: Conference Series **199** (2010) 012016 doi:10.1088/1742-6596/199/1/012016

and Research for support under the International Linkages Program. In particular we are grateful to our colleagues M. Charlton and C. Surko for useful comments and discussions related to this work.

References

[1] Marler JP, Sullivan JP and Surko CM 2005 *Phys. Rev. A* **71** 022701

[2] Laricchia G, Van Reeth P, Szluinska M and Moxom J 2002 *J. Phys B.* **35** 2525

[3] Marler JP and Surko CM 2005 *Phys. Rev. A* **72** 062702

[4] Marler JP and Surko CM 2005 *Phys. Rev. A* **72** 062713

[5] Charlton M 1985 *Rep. Prog. Phys.* **48** 737

[6] Gilmore S, Blackwood JE and Walters HRJ 2004 *Nucl. Instr. Meth. B* **221** 129

[7] Baluja K L, Rui Zhang, Franz J. and Tennyson J 2007 J. Phys. B: At. Mol. Opt. Phys. **40** 3515

[8] Campeanu RI 1982 *Canadian Journal of Physics*, **60**,615-624.

[9] Charlton M 2009 in Nonequilibrium processes in plasma physcs *J.Phys. Conference Series*,

[10] Bose N., Paul D. A. L., and Tsai J. S., 1981, J. Phys. B 14, L227.

[11] Charlton M 1985 J. Phys. B: At. Mol. Phys. 18 L667

[12] Shizgal B, Ness KF, 1987 J. Phys. B: Atom. Molec. Phys. **20**, 847.

[13] Charlton M and Humberston J W 2001 *Positron Physics* (Cambridge University Press)

[14] Shakeshaft R, *Wadhera* JM, 1980 *Phys. Rev.* A. **22**: 968

[15] Gilbert S. J., C. Kurz, R. G. Greaves, and C. M. Surko, Appl. Phys. Lett. 70, 1944 (1997).

[16] Robson R E 1991 Aust. J. Phys. 50 577

[17] Šuvakov M, Petrović ZLj, Marler JP, Buckman SJ, Robson RE and Malović G 2008 *New J. Phys.* **10 No 5** 053034

[17] Marler JP, Petrović ZLj, Banković A, Dujko S, Šuvakov M, Malović G and Buckman SJ 2009 *Phys. Plasmas* **16**, 057101

[18] Banković A, Marler JP, Šuvakov M, Malović G and Petrović ZLj 2008 *Nuclear Inst. Methods, B* **266** 462

[19] Vrhovac SB and Petrović ZLj 1996 *Phys. Rev. E* **53** 4012

[20] White and Robson 2009 *Phys. Rev. Lett.* **102** 230602 and unpublished

[21] Banković A, Dujko, Malović G, Buckman SJ, White RD, Marler JP, Petrović ZLj 2009 unpublished.

[22] Al-Qaradawi I, Charlton M, Borozan I and Whitehead R. 2000 *J. Phys. B: At. Mol. Opt. Phys.* **33** 2725

[23] S. Marjanović, M Šuvakov, A Banković, S Dujko, SJ Buckman, JP Marler, ZLj Petrović 2009 to be published

[24] R. G. Greaves and C. M. Surko, 2000, *Physical Review Letters* **85,** 1883

[25] J. J. Engbrecht, M. J. Erickson, C. P. Johnson, A. J. Kolan, A. E. Legard, S. P. Lund, M. J. Nyflot, and J. D. Paulsen *Phys. Rev. A* **77**, 012711 (2008)

[26] S. Marjanović, Šuvakov M, Engbrecht JJ, Petrović ZLj 2009 to be published

Journal of Physics: Conference Series **199** (2010) 012017 doi:10.1088/1742-6596/199/1/012017

Theoretical calculation of total cross sections for e^+ - NH_3 molecule at low energies

Arvind Kumar Jain[1], Indira Dubey[2] and Harsh Mohan[1]

[1]Physics Department, M.L.N. College, Yamuna Nagar 135 001, Haryana, India
[2]Laser Cosmetics and Pain Treatment Clinic, 700 Balmoral Dr., #7 Brampton, ON L6T 1X2, Canada

e-mail: arvindjain61@gmail.com

Abstract. A parameter-free spherical complex optical potential model is used to calculate the differential and total cross sections for the scattering of positrons by the NH_3 molecules in the energy range 10 eV - 400 eV. The optical potential is constructed from a near-Hartree-Fock one-center expansion of ammonia (NH_3) wave function and it is then treated exactly in a partial-wave analysis involving variable phase approach to extract complex phase shifts. The present theoretically calculated results show good agreement with the available experimental results for total cross sections at positron energy $\geq$ 30 eV.

1. Introduction

The total cross sections (including elastic plus energetically possible all inelastic channels) for electron/positron collisions with a large number of atoms and polyatomic molecules have been experimentally measured from as low as (~ 1 eV) to keV energy region [1-3]. Theoretical description of positron scattering from molecules is given by Armour [4]. These cross sections have important applications in space, plasma, laser, biomedical science and the environment. In particular, the discovery of NH_3 molecules in the interstellar medium has increased the attention paid to its spectroscopic and dynamical properties [5], whereas in the field of plasma chemistry, NH_3 has gained importance as it is a source of nitrogen atoms for the fabrication of nitride films and the synthesis of other nitrogen compounds [6]. On the theoretical side, to study the scattering of electrons and positrons with diatomic and polyatomic molecules, many approaches ranging from as simple as independent atom model (IAM) [7] to *ab initio* methods [8] have been proposed and developed. In case of positrons colliding with ammonia molecules, IAM approach was used by Raizada and Baluja [9]. In their calculations, total cross section for positron impact on a molecule was expressed as an incoherent sum over the total cross sections of the atoms of the molecules. The rule is called as the additivity rule. It has also been applied for the electron impact on the various molecules by many workers [10-12]. On the experimental side, total cross sections for positrons and electrons colliding with NH_3 molecules were measured by Sueoka *et al* [13] using time-of-flight (TOF) method in the energy range 1 – 400 eV. They had applied a magnetic field of 4.5 G in the range of 1 – 7.5 eV, of 9 G in the range of 0.7 – 400 eV in their experimental technique. The experimental cross sections were obtained by a normalization method. As *Ab initio* calculations are quite difficult to perform in case of polyatomic molecules, therefore, here our goal is to apply a spherical complex optical potential (SCOP) approach for calculating differential cross sections (DCS) and total (σ_t) for positrons colliding with NH_3 molecules in the energy range 10 – 400 eV.

2. Methodology

The total interaction of the positron molecule system can be represented by a local complex optical potential, namely,

$$V_{opt}(\boldsymbol{r}) = V_R(\boldsymbol{r}) + i\, V_{Abs}(\boldsymbol{r}) \tag{1}$$

where the real part $V_R(\boldsymbol{r})$ of the optical model potential, consists of the static potential V_{st} and parameter free correlation polarization potential V_{cop}. The complex optical potential is obtained by adding a model absorption potential $V_{Abs}(\boldsymbol{r})$ as imaginary part to the real potential. All the potentials can be generated easily once the target charge density $\rho_0(\boldsymbol{r})$ is known.

The ammonia molecule belongs to the $C_{3\upsilon}$ symmetry point group (electronic $^1A_1$ ground state) with the electronic configuration $1a_1^2 2a_1^2 1e^4 3a_1^2$. A single-center expansion technique (obtained from Jain and Thompson [14]) with a nitrogen atom at the center is employed for all orbitals in the near-Hartree-Fock limit using experimental values for nuclear geometry (such as bond length 1.012Å and bond angle = 106.7°). We determine $\rho_0(\boldsymbol{r})$ with enough terms in the expansion of each orbital. $\rho_0(\boldsymbol{r})$ is expanded in terms of symmetry-adapted functions belonging to the totally symmetric $^1A_1$ irreducible representation of the molecular $C_{3\upsilon}$ point group

$$\rho_0(\boldsymbol{r}) = \sum \bar{\rho}_{LH}(\boldsymbol{r})\, X_{LH}^{A_1}(\hat{r}) \tag{2}$$

In the spherical approximation, we only need the first term ($L = 0$, $H = 1$) of the expansion of equation (2), in order to evaluate all the three terms, i.e. V_{st}, V_{cop} and V_{Abs}. Explicit expression for V_{st} is given in the literature (see, for example, Gianturco and Jain [15]).

The attractive term i.e. polarization effect, which arises due to distortion in the target charge cloud because of incoming charge particle. This effect is invoked by a parameter free correlation polarization potential, which is based on the correlation energy of the target molecule. The calculation of an actual polarization potential which provides the second and higher order effects in the positron - molecule interaction is rather a difficult task. Following Padial and Norcross [16] and Gianturco *et al* [17] the correlation polarization potential, for various range of radial distance (r) is expressed in terms of the density parameter r_s

$$r_s(\boldsymbol{r}) = \left[\frac{3}{4\pi\rho_0(\boldsymbol{r})}\right]^{1/3}$$

as,

$$V_{cop}(\boldsymbol{r}) = \begin{cases} 0.0311\, ln\, r_s + 0.006\, r_s\, ln\, r_s - 0.015\, r_s - 0.0584, & r_s \le 0.7 \\ 0.02224\, ln\, r_s - 0.07356, & 0.7 \le r_s \le 10.0 \\ -0.584\, r_s^{-1} + 1.988\, r_s^{-3/2} - 2.45\, r_s^{-2} - 0.733\, r_s^{-5/2}, & 10.0 \le r_s \end{cases} \tag{3}$$

The imaginary part of the optical model potential is referred as absorption potential, which takes into account the flux going through all possible inelastic channels. An exact determination of V_{abs} is a very difficult task. Thus a number of empirical and semi-empirical models have been proposed. In the present work, it is determined *ab initio* as the function of charge density of the target molecule following the procedure of Staszewska *et al* [18]. This is written as

$$V_{abs}(r) = -\rho_0(r)(v_{loc}/2)^{\frac{1}{2}}\left(\frac{8\pi}{10k_f^3E}\right)X\, H(k^2 - k_f^2 + 2\Delta)(A_1 + A_2 + A_3) \tag{4}$$

Where,

$$v_{loc} = \frac{1}{2}k^2 - V_{st} - V_{pol}$$

$$A_1 = 5k_f^3/2\Delta$$

$$A_2 = -k_f^3(5k^2 - 3k_f^2)/(k^2 - k_f^2)^2$$

$$A_3 = 2H(2k_f^2 + 2\Delta - k^2)\, X\, [(2k_f^2 + 2\Delta - k^2)^{5/2}/(k^2 - k_f^2)^2]$$

Here *H(x)* is a Heaviside function defined by

$$H(x) = \begin{cases} 1, & x \geq 0 \\ 0, & x < 0 \end{cases}$$

where Δ is the mean excitation energy which is obtained in the closure approximation from the ground-state target wave function ψ_0 of the target molecule.

$$\Delta = \frac{2\langle \psi_0 | z^2 | \psi_0 \rangle}{\alpha_0}$$

For NH_3, Δ is calculated to be 14.986 eV.

Finally, a partial wave analysis approach is adopted to compute the ℓth partial wave phase shifts for the solution of radial Schrodinger equation and a variable-phase approach (VPA) [19] and has been employed to find its solution. The corresponding quantities (i.e. DCS and σ_t) are then easily obtained from S-matrix at each energy. All our cross sections are converged with respect to the number of partial waves up to a value of 0.001 radians only.

3. Results and discussion

In figure 1, DCS curves have been shown at incident energy 50 eV. At this energy, neither theoretical calculation nor experimental measurements for positrons scattering with ammonia molecule is available in the literature. Therefore, we have calculated DCS by using both complex and real potential in order to study the influence of the absorption potential. Noticeably, in the present fixed-nuclei approximation the DCS's at $\theta = 0^{\circ}$ are not defined. It is clear from the figure that the absence of absorption potential reduces the DCS in magnitude, which is characterized by a dip in the angular distribution at the angles 50° and 65°.

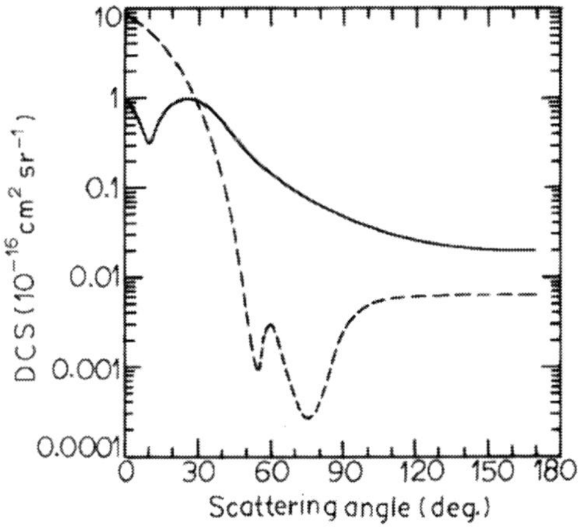

Figure 1. Differential cross sections for e^+-NH_3 scattering. Present calculations: _____, with absorption; -----, without absorption.

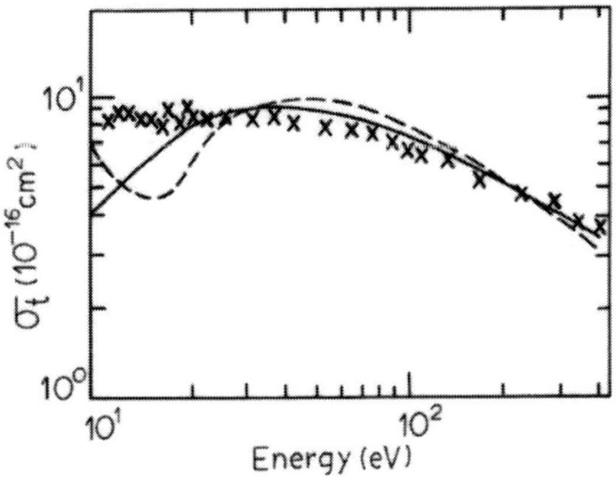

Figure 2. Total cross-sections (σ_t) for e^+ - NH_3 scattering. Calculations: _____, present results; ----, Raizada and Baluja [9]. Experiment: X, Sueoka *et al* [13].

Total cross sections are also evaluated from the DCS's. Further, one can avoid the forward angle-peaking behavior of the DCS's by calculating the corresponding momentum-transfer cross section due to the $(1 - \cos\theta)$ term. In figure 2, the σ_t cross sections for e^+ - NH_3 are presented and compared with the IAM calculations of Raizada and Baluja [9] together with experimental measurements of Sueoka *et al* [13]. It is observed that our theoretical results are in better agreement with the experimental data at all energies except below 30 eV. At lower energies, the disagreement may be due to the contribution of the dipole, quadrupole etc. terms, which plays an important role in non-spherical targets. The comparison between our results and the IAM calculations [9] points out that the effect of bonding is dominant factor in this energy region.

4. Conclusions

It is finally concluded that a parameter-free model potential constructed for the full interaction of the collision system within the framework of the spherical-complex-optical model potential approach is quite adequate to yield the differential and total cross sections. The calculated total cross sections are in good agreement with the available experimental measurements.

Acknowledgement

It is a matter of pleasure to acknowledge the financial support for carrying out this research work from Government of India through the Department of Science and Technology (DST), New Delhi, India under the project no. SR/S2/LOP-0018/2006.

References

[1] Buckman S J, Maddern T, Francis-Staite J, Hargreaves L, Brunger M J, Garcia G, Lower J C, Mondal S, Sullivan J P, Jones A, Caradonna P, Slaughter D, Mackochekanwa C and McEachran R P 2008 *J. Phys.: Conference Series* **133** 012001
[2] Surko C M, Gribakin G F and Buckman S J 2005 *J. Phys. B: At. Mol. Opt. Phys.* **38** R57
[3] Jain A 1987 *Positron Annihilation Studies of Fluids* ed S C Sharma and D M Schrader (Singapore: World Scientific)
[4] Armour E A G 1988 *Phys. Rep.* **169** 198
[5] Gianturco 1991 *J. Phys. B: At. Mol. Opt. Phys.* **24** 4627
[6] Sato T, Shibata F and Goto T 1986 *Chem. Phys.* **108** 147
[7] Jiang Y, Sun J and Wan L1995 *Phys. Rev.* A **52** 398
[8] Gianturco F A, Lucchese R R and Sanna N 1994 *J. Chem. Phys.* **100** 6464. See also Natalense A P P, Beltega H H F, Ferreira L G and Lima H A P 1995 *Phys. Rev.* A **52** R1
[9] Raizada R and Baluja K L 1997 *Phys. Rev.* A **55** 1533
[10] Joshipura K N and Patel P M 1994 *Z. Phys.* D **29** 269
[11] Younger S M and Mark T D 1985 *Electron Impact Ionization* vol. 1 (Berlin: Springer)
[12] Baluja K L and Agarwal A 1995 *Phys. Lett.* **198** 225
[13] Sueoka O, Mori S and Katayama K 1987 *J. Phys.* B **20** 3237
[14] Jain A and Thompson D G 1987 *J. Phys. B: At. Mol. Phys.* **20** 2861
[15] Gianturco F A and Jain A 1986 *Phys. Rep.* **143** 347
[16] Padial N T and Norcross D W 1984 *Phys. Rev.* A **29** 1590
[17] Gianturco F A, Jain A and Pantano L C 1987 *J. Phys.* B **20** 571
[18] Staszewska G, Schwenke D, Thirumalai D and Truhlar D G 1983 *J. Phys.* B **16** L281
[19] Calagero F 1974 *Variable Phase Approach to potential Scattering* (New York: Academic)

Ionization of the water molecule by electron and positron impact

I. Tóth[1], **R. I. Campeanu**[2], **V. Chiş**[1] **and L. Nagy**[1]

[1] Faculty of Physics, Babeş-Bolyai University, Kogălniceanu 1, 400084 Cluj Napoca, Romania
[2] Department of Physics and Astronomy, York University, 4700 Keele Street, Toronto, Canada

E-mail: itoth@phys.ubbcluj.ro

Abstract. Theoretical cross-sections for the ionization of the water molecule by electron and positron impact are presented. The calculations were performed in the framework of the simpler CPE and two distorted-wave models, ES and TS, by employing Gaussian wavefunctions for the description of the target. We found good agreement with the experiment, especially for higher impact energies.

1. Introduction

Water has a large practical importance in collision processes with positron and electron beams. Because a large percentage of living tissue is water, all medical diagnosis and treatment methods need precise cross section data for the water molecule in collision with charged particles. These collisions play an important role also from the viewpoint of other research fields like astrophysics or chemistry.

The interaction of water with positron and electron projectiles, has received significant attention in the last years. Both experimental [1, 2, 3] and theoretical [4, 5, 6], total cross-sections have been reported for the positron-water scattering process. Other experiments have measured positronium-water total cross sections [3] and even double-differential cross-sections [7] for the ionization of H_2O by positron impact. Unfortunately, there are no experimental data for the direct ionization of the water molecule by positron impact. For electron impact ionization, experimental data [8, 9, 10, 11] exist even for the direct ionization process.

Previously we have calculated cross-sections for the direct ionization of several molecules by positron impact, like H_2 [12], N_2 [13], O_2 [14], CO, CO_2 [15], CH_4 [16]. Later, we have refined [17] and extended [18] our calculations to the ionization of the studied molecules by electron impact. These studies were based on the simple CPE (Coulomb plus plane waves with full energy range) and the more complex distorted-wave models, ES (Electron Screening) and TS (Total Screening). In each case, the initial state of the active electron was described by Gaussian-type wavefunctions.

In this paper we extend our calculations to the direct ionization of the water by electron and positron projectiles. The obtained results are compared with the existent experimental data [8, 9, 10, 11], the theoretical results of [19], and the DWBA (Distorted-Wave Born Approximation) calculations of Champion et al. [20] in the case of electron impact ionization. Champion et al. [20] have employed the Slater-type functions of Moccia [21] for the description of the initial state of the target molecule. They have considered each molecular orbital as a linear combination of Slater-type functions, all centered at a common origin, i. e. the oxygen atom. They have employed plane waves for the description of the incident and scattered particles, while the ejected electron was represented by a distorted-wave.

2. Theory

The present calculation for the water molecule follows closely the method employed in the case of the CH_4 [16], since both are nonlinear molecules. Here, we give a brief overview of the main steps of the theory.

The triple differential cross-section for the ionization of a molecular target by electron or positron impact may be written as

$$\frac{d^3\sigma}{d\hat{\mathbf{k}}_f d\hat{\mathbf{k}}_e dE_e} = \sum_r \frac{(2\pi)^4}{E_i} |f_t^r|^2 , \qquad (1)$$

where the sum runs over all occupied molecular orbitals. In (1) the continuum wave functions were normalized in energy, and the energy is expressed in Rydbergs. E_i denotes the impact energy, while E_e stands for the energy of the ejected electron. $\hat{\mathbf{k}}_f$ and $\hat{\mathbf{k}}_e$ stand for the angles of the momenta of the scattered and ejected particles, respectively. For a positron projectile the total scattering amplitude, f_t^r, coincides with the f_d^r direct amplitude, while for electron projectiles the inclusion of the exchange effects lead to

$$|f_t^r|^2 = |f_d^r|^2 + |f_{ex}^r|^2 - \alpha\, |f_d^r|\, |f_{ex}^r| . \qquad (2)$$

Here f_{ex}^r is the exchange amplitude, while α is a phase factor [22]. The direct and the exchange amplitudes may be written as

$$f_d^r = \langle \phi_f(\mathbf{r}_1)\phi_e(\mathbf{r}_2)\, |V(r_{12})|\, \phi_i(\mathbf{r}_1)\phi_b^r(\mathbf{r}_2)\rangle \qquad (3)$$

$$f_{ex}^r = \langle \phi_e(\mathbf{r}_1)\phi_f(\mathbf{r}_2)\, |V(r_{12})|\, \phi_i(\mathbf{r}_1)\phi_b^r(\mathbf{r}_2)\rangle, \qquad (4)$$

where ϕ_i, ϕ_e and ϕ_f are the wave functions of the incident, ejected and scattered particles, respectively. In order to make the calculations tractable, these wave functions were expanded in terms of the partial waves. In the direct amplitude $\mathbf{r}_1$ is the position vector of the projectile and $\mathbf{r}_2$ gives the position of the active electron, while in the exchange amplitude the states of the scattered and the ejected particles are exchanged. ϕ_b^r stands for the initial state (orbital) of the active electron. We have taken into account the following orbitals of the water molecule: $2a_1$, $1b_2$, $3a_1$ and $1b_1$ with the corresponding ionization potentials (in atomic units): 1.26805, 0.61744, 0.45300 and 0.39122. The molecular orbitals are constructed as linear combination of the O and the two H Gaussian orbitals:

$$\phi_b^r(x, y, z) = C_O^r \phi_O^r(x, y, z) + \sum_{i=1}^{2} C_{H_i}^r \phi_{H_i}^r(x - x_i, y - y_i, z - z_i). \qquad (5)$$

Here, x_i, y_i and z_i are the coordinates of the hydrogen atoms. We have obtained the Gaussian orbitals as contractions of Gaussian-type functions. As in the case of the methane molecule [16], the multi-centre molecular orbitals were expanded in terms of the spherical harmonics

$$\phi_b^r(\mathbf{r}_2) = \sum_{l_b\mu} C_{l_b\mu}^r(r_2) Y_{l_b\mu}(\hat{\mathbf{r}}_2), \qquad (6)$$

where the expansion coefficients may be expressed as

$$C_{l_b\mu}^r = \int d\hat{\mathbf{r}}_2 Y_{l_b\mu}(\hat{\mathbf{r}}_2) \phi_b^r(\mathbf{r}_2). \qquad (7)$$

In order to calculate the ionization cross section, one has to integrate the differential cross-section (1) over the angles of the scattered and ejected electron as well as the energy of the ejected electron. Further, the cross-section is averaged over all molecular orientations.

We have calculated the cross-sections for the ionization of the H_2O molecule in the framework of three models developed earlier, the CPE, ES and TS models. The CPE [23] is a simple model, which uses Coulomb or plane waves for the description of the free particles. The ES and TS are distorted-wave models. The distorted-waves were calculated by solving the radial Schrödinger equation in the field of a spherically averaged potential created by the nuclei and the electrons of the target. The ES model employs the distorted-wave description for the ejected electron, while the TS model uses the distorted-wave formalism for the description of all free particles. Details about the calculational method and these models were given in papers [17] and [18].

3. Results and discussions

We present the results for the direct ionization of the H_2O molecule in Figures 1-2 and Tables 1-2, respectively.

Fig. 1 shows the integrated cross-sections for the direct ionization of the water molecule by electron impact, determined in the framework of the CPE, ES and TS models. Our calculations are compared with various experimental data [8, 9, 10, 11] and the theoretical results [19] and [20].

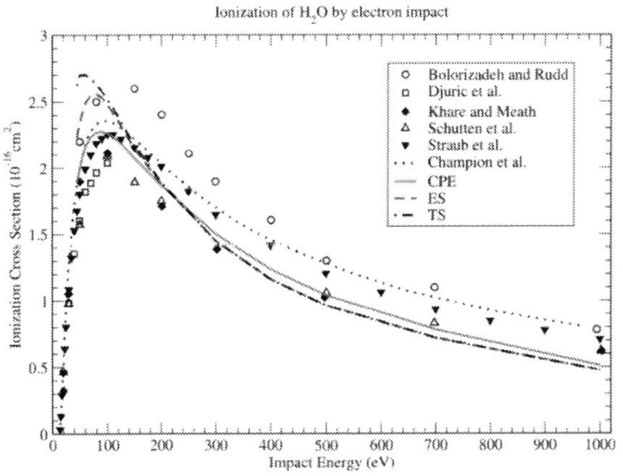

Figure 1. Cross-sections for the direct ionization of the H_2O molecule by electron impact, calculated in the framework of the CPE, ES and TS models, compared with the measurements of [8, 9, 10, 11] and the theoretical results [19] and [20].

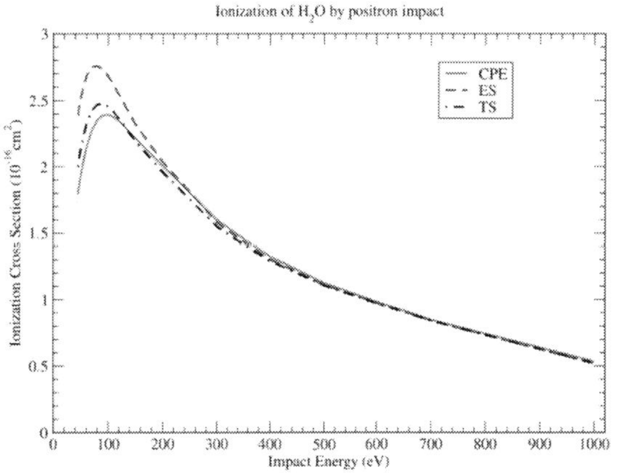

Figure 2. Cross-sections for the direct ionization of the H_2O molecule by positron impact, calculated in the framework of the CPE, ES and TS models. At the date of the study there were no experimental data for the direct ionization of this target.

It may be observed that our CPE model provides a good agreement with the measurements of Schutten et al. [10] and the results of Khare and Meath [19], respectively. The agreement is especially good for impact energies higher than 200 eV. In this energy range our results are slightly lower than the measurements of Bolorizadeh and Rudd [8] and Straub et al. [11]. The calculated cross-section of Champion et al. [20] agree well with the measurements of Straub et al. [11], but is higher than the cross-sections of [9, 10, 19]. In the region of the peak, the CPE results are a slightly higher than the measurements of Djuric et al. [9], Schutten et al. [10] and

Journal of Physics: Conference Series **199** (2010) 012018 doi:10.1088/1742-6596/199/1/012018

the calculated cross-sections of Khare and Meath [19], but still are in reliable agreement with the data of Straub et al. [11]. However, the peak of the CPE model is slightly shifted to lower energies. The ES and TS models provide good agreement with the data of [10] and [19] but overestimate these results at the region of the peak. It seems that the approximation introduced by spherically averaging the real potential created by the nuclei and electrons of the target is not suitable in this case, where the CPE model, which uses simple Coulomb potentials, provide better results. For energies higher than 200 eV the distorted-wave models give lower values for the cross-section than the measured data of [8] and [11] and the theoretical calculations of [20].

The results obtained within the CPE, ES and TS models for the ionization of the water molecule by positron projectiles are presented in Fig. 2. There are no experimental data for the direct ionization process of this target. The obtained cross-sections for electron and positron projectiles show similar values. Nevertheless, the cross-sections for positron impact are slightly higher than for electron projectiles, except for the TS model. In the region of the peak, the distorted-wave models ES and TS show higher values than the CPE model, similarly as in the case of the electron cross-section. For impact energies higher than 200 eV the positron cross-sections calculated in the framework of the different models seem to merge. It will be interesting to compare these results with the experimental measurements expected in the near future [24].

An analysis of the contributions to the total cross section of the different orbitals taken into account and the contributions coming from the different terms of the expansion on l_b is provided in Table 1 for electron projectiles and in Table 2 for positron projectiles.

Table 1. Partial ionization cross-sections (in 10^{-16} cm^2) calculated in the framework of the ES model for an electron projectile of 50 eV impact energy.

	l_b				
Orbital	0	1	2	3	Total Orbital
$2a_1$	9.67×10^{-2}	4.49×10^{-4}	7.47×10^{-5}	3.36×10^{-5}	0.097
$1b_2$	0	0.64	1.07×10^{-2}	6.52×10^{-4}	0.651
$3a_1$	0.10	0.22	7.22×10^{-3}	6.10×10^{-4}	0.327
$1b_1$	0	1.24	9.08×10^{-3}	1.89×10^{-4}	1.249
Total l_b	0.196	2.100	0.027	1.48×10^{-3}	2.324

Table 2. Partial ionization cross-sections (in 10^{-16} cm^2) calculated in the framework of the ES model for a positron projectile of 50 eV impact energy.

	l_b				
Orbital	0	1	2	3	Total Orbital
$2a_1$	7.71×10^{-2}	4.98×10^{-4}	9.51×10^{-5}	2.94×10^{-5}	0.077
$1b_2$	0	0.750	1.18×10^{-2}	8.24×10^{-4}	0.762
$3a_1$	0.106	0.246	7.68×10^{-3}	6.42×10^{-4}	0.360
$1b_1$	0	1.308	1.23×10^{-2}	1.03×10^{-4}	1.320
Total l_b	0.183	2.304	0.031	0.001	2.52

The tables list the contributions for an impact energy of 50 eV, in the case of the ES model. In each case, for large l_b values the contributions are less important than for smaller l_b values. This is the reason why we have neglected the $l_b > 3$ terms. As Tables 1 and 2 show, the first molecular orbital ($2a_1$) taken into account has mainly a spherical character (the $l_b = 0$ term is dominant), the contributions from higher order terms are negligible. In the case of the $1b_2$ and the $1b_1$ orbitals the $l_b = 1$ term is dominant, therefore the orbitals show a mainly dipole character. For the $3a_1$ orbital, both the monopole ($l_b = 0$) and dipole ($l_b = 1$) terms contribute significantly to the cross section.

4. Conclusions

The ionization of the water molecule by electron and positron impact was investigated. We found a fair agreement between the experimental data and our results. Particularly, the good agreement with the measurements provided by the simpler CPE model is impressive. It seems that in the region of the peak, our distorted-wave models overestimate the experimental results for electron impact ionization. Nevertheless, for higher impact energies the agreement is good. The cause of the discrepancy may be the spherical nature of the potential employed to calculate the distorted-waves.

Although there are no measurements for the ionization of the water by positron impact, we found similar values for positron and electron cross-sections. In both cases, our distorted-wave models provide higher values than the CPE in the region of the peak. However, the positron cross-sections are slightly higher than for electron cross-sections, except for the TS model.

Acknowledgments

The authors acknowledge the support of the Romanian National Plan for Research (PN II) under contract No. ID_539 and by a grant from the Natural Sciences and Engineering Research Council of Canada.

References

[1] Sueoka O, Mori S and Katayama Y 1986, *J. Phys. B: At. Mol. Phys* **19** L373
[2] Zecca A, Sanyal D, Chakrabarti M and Brunger M J 2006, *J. Phys. B: At. Mol. Phys* **39** 1597
[3] Beale J, Armitage S and Laricchia G 2006, *J. Phys. B: At. Mol. Opt. Phys.* **39** 1337
[4] Baluja K L and Jain A 1992, *Phys. Rev.* A **45** 7838
[5] Gianturco F A, Mukherjee T and Occhigrossi A 2001, *Phys. Rev.* A **64** 032715
[6] Baluja K L, Zhang R, Franz J and Tennyson J 2007, *J. Phys. B: At. Mol. Opt. Phys.* **40** 3515
[7] Arcidiacono C, Kövér Á, Pesic Z D, Beale J and Laricchia G 2007, *Rad. Phys. Chem.* **76** 431
[8] Bolorizadeh M A and Rudd M E 1985 *Phys. Rev.* A **33** 882
[9] Djuric N Lj, Cadez I M and Kurepa M V 1988 *Int. J. Mass Spectrom. Ion Processes* **83** R7
[10] Schutten J, de Heer F J, Moustafa H R, Boerboom A J H and Kistenmaker J 1966 *J. Chem. Phys.* **44** 3924
[11] Straub H C, Lindsay B G, Smith K A and Stebbings R F 1998 *J. Chem. Phys.* **108** 109
[12] Campeanu R I, Chiş V, Nagy L and Stauffer A D 2003, *Phys. Lett.* A **310** 445
[13] Campeanu R I, Chiş V, Nagy L and Stauffer A D 2004, *Nucl. Instrum. Methods* B **221** 21
[14] Campeanu R I, Chiş V, Nagy L and Stauffer A D 2004, *Phys. Lett.* A **325** 66
[15] Campeanu R I, Chiş V, Nagy L and Stauffer A D 2005, *Phys. Lett.* A **344** 247
[16] Campeanu R I, Chiş V, Nagy L and Stauffer A D 2006, *Nucl. Instrum. Methods* B **247** 58
[17] Tóth I, Campeanu R I, Chiş V and Nagy L 2006, *Phys. Lett.* A **360** 131
[18] Tóth I, Campeanu R I, Chiş V and Nagy L 2008, *Eur. Phys. J.* D **48** 351
[19] Khare S P and Meath W J 1987 *J. Phys.* B **20** 2101
[20] Champion C, Hanssen J and Hervieux P A 2002, *J. Chem. Phys.* **117** 197
[21] Moccia R 1964, *J. Chem. Phys.* **40** 2186
[22] Younger S M 1980, *Phys. Rev.* A **22** 111
[23] Campeanu R I, McEachran R P and Stauffer A D 1987, *J. Phys. B: At. Mol. Opt. Phys.* **20** 1635
[24] Laricchia G and Sullivan J 2009, *private communications*

Energy scale determination and energy resolution in positron total cross section measurements

G P Karwasz[1], R S Brusa[2] and D Pliszka[3]
[1] Institute of Physics, Nicolaus Copernicus University, 87-100 Toruń, Poland
[2] Dipartimento di Fisica, Università di Trento, 38010 Trento, Italy
[3] Institute of Physics, Pomeranian Academy, 76200 Słupsk, Poland

E-mail: karwasz@fizyka.umk.pl

Abstract. We compare the recent values of positron total cross sections in benzene, cyclohexane and aniline by Zecca et al. with our previous measurements [Karwasz et al. Acta Phys. Pol. A, 107 (2005) 666] on the same apparatus. An apparent discrepancy disappears if Zecca's data are shifted by +0.2 eV; then their data can be approximated by the modified effective range theory, as well. This observation induced us to compare recent Trento data with other experiments and theories for polar targets, H_2O and HCOOH. The evidence points out towards the need of a careful energy calibration in positron experiments.

1. Introduction

Measurements of total cross sections rarely extend down to sub-eV energy range. This was, for example, the case of the Wayne State University (Detroit) experiment in its early version, using boron ^{11}B radioisotope what allowed to perform measurements of total cross sections in Ar [1] down to as little 0.4 eV while the measurements with a thick (ribbon) W moderator stopped at 4 eV [2]. Recently a new version of the tungsten moderator a thin 1 μm micro foil has been developed at Trento University [3-6], allowing measurements of total cross section down to a few tenths of eV.

Two distinct groups performed measurements on the Trento apparatus [5] – present authors [7-10], and more recently Zecca and collaborators who published data on simple targets as H_2O [11] and on more complex, organic molecules [12-14]. The two sets from Trento apparatus for targets like benzene and aniline apparently differ. The newest-type moderator i.e. the solid Ne, followed by a gas-cell re-moderation stage can yield as good as 25 meV energy resolution [15]. It is currently used at Australian National University in Canberra with the energy resolution of 65 meV [16] and some preliminary data were shown at the present Workshop.

The aim of the present paper is to show some data coming from those recent experiments and compare them with the most recent theories [17]. In particular we discuss the energy scale as used in Zecca's *et al.* reports [11-14]. We show that for all the discussed targets - atoms (He, Ar), hydrocarbons (C_6H_6, C_6H_{12}, C_6H_5N) and polar molecules (H_2O, HCOOH) there exist a very good agreement between the recent theories and (both) sets of Trento experimental results, once a proper energy scale is given. In detail, the analysis shows that the energy calibration in Zecca's data has to be adjusted by some +0.2 eV.

2. Energy scaling procedures

Measurements of electron (and positron) scattering are subject to the energy-bias uncertainty mainly due to electrochemical contact potentials between the source and the scattering cell. The most objective way of determining the energy scale shift are direct measurements of some sharp structures in cross sections, like the resonant vibrational structure in N_2 [18] or sharp resonances in He [19] or Ar [20]. For positrons any sharp structures in cross sections are still a subject of search [21]. In their absence some substituting measures have to be found. The use of electron cross sections, like the N_2 resonant vibrational structure, is one of them. As seen from figure 1, the Trento apparatus with the electron optics set for positron measurements is able to reproduce that structure, and the deduced energy resolution is about 150 meV.

Unfortunately, the methods mentioned above can not be used to determine the energy scale. The work function of positron re-emitted from the tungsten surface is positive while for electrons it is negative. In our apparatus neither the time-of-flight calibration method [15, 22] can be used. Therefore, we have calibrated the energy scale against sharp features like the positronium-formation thresholds in N_2 and Ar [8] which are known with a high accuracy. In particular, in Ar no (rotational, vibrational or electronic) excitations are possible below the Ps threshold (8.96 eV) so the onset is well visible, see figure 2. Furthermore, as seen in our data for Ar, as noted by Szłuińska and Laricchia for Xe [23] and as recently confirmed by Buckman and collaborators in Ar [24], threshold structures are present near the positronium threshold. From Ar measurements we deduced the potential bias of +2.4 eV between the W moderator and the scattering cell. Such a value is in a perfect agreement with measurements of the energy spectrum for positrons emitted from the same-type moderator [25].

The spectrum of positrons emitted from W-moderator is significantly different from that for secondary electrons which extends from zero up to some tens of eV. The Trento apparatus is not equipped with an energy-selector in sensu stricte, see details in figure 1 of Ref. [8]. Three elements perform, however, the energy selecting function: 1) the positron source, yielding a few-eV wide spectrum, 2) the deflection bent, with 1.2 eV FWHM pass-band and 3) the longitudinal magnetic field in the scattering cell. In particular, that latter, as proved in constructing a similar drift-tube [26] for the positron microscope [3], assures the transmission only for positrons at strictly quantified energies (say 2, 4, 6 eV etc.). These three elements perform de facto the monochromatic action on positron, but not on the secondary electrons. For electrons, in particular at low energies, more than one quantified energy can be transmitted through the scattering cell. In fact, as seen from the comparison of Zecca's *et al.* data in tetrohydrofuran [11] with other laboratories [27, 28], his electron data are in a serious error; as a consequence, Zecca has shown no electron data for successive targets.

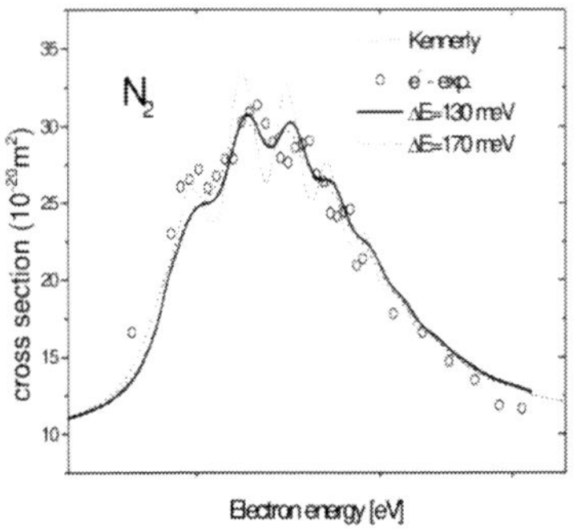

Figure 1. Performance tests (energy resolution evaluation) of Trento [8] apparatus - vibrational resonant structure in N_2 electron-scattering [18] compared to the structure measured with Trento positron-scattering apparatus [8]. The measured structure indicates 130-150 meV energy resolution.

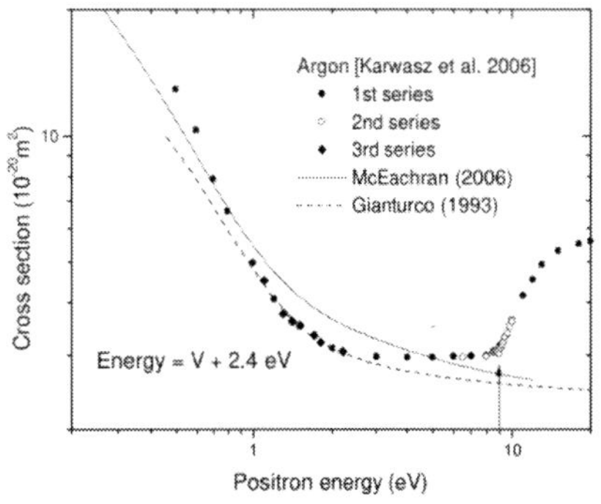

Figure 2. Performance tests (energy calibration) of Trento [8] apparatus for argon. Points – Karwasz et al. [8], separate experimental series. Theory: broken line, Gianturco et al. [29]; solid line, McEachran and Stauffer [30]. The arrow shows the positronium - formation threshold.

3. Comparison of experimental and theoretical data

Comparison of different sets of experimental (and theoretical) data is a supplementary way of determining the construction features of the apparati. For argon we compare our experimental data [8] with two recent theories - by Gianturco *et al.* [29] polarisation potential from free-electron-gas derivation with dipole potential and by McEachran [30] using the complex ab-initio optical potential. The two theories agree pretty well with the experiment and the modified effective-range fit [31]. However, only the theory by Gribakin and Ludlow [32] using many-body theory with perturbation corrections to the correlation potential is able to reproduce the constant value of the cross section between about 2 eV and the positronium-formation threshold. Some spread between theories still does not allow to make decisive judgments and more measurements are needed.

In figure 3 we compare present experiment for benzene with those by Zecca *et al.* [14] and the early data by Sueoka *et al.* [21]; we do not show comparison with some more recent experiments from Tokyo lab, as they were, probably, subject to a big angular-resolution error [33]. Data of Zecca *et al.* almost coincide with ours, but they must be shifted by +0.2 eV. After this shifting, all three sets agree very well up to the positronium-formation threshold (signed by the arrow in fig. 3). Above the positronium-formation threshold the data of Sueoka [22] deviate downwards. This can be explained by opening of this new channel: for rather large (3 mm radius) exit apertures of the scattering cell, the electrically neutral positronium, if formed in the forward direction, can reach the detector and lower the measured cross section. This effect can be superimposed on the confining action of rising values of the magnetic field used at higher energies by Sueoka. Zecca's *et al.* [14] presently corrected values are approximated well also by the modified effective range theory [34].

Also for cyclohexane and aniline, see figure 4, the data by Zecca *et al.* [14], after the energy correction by +0.2 eV, coincide with our results [7]. Present data in cyclohexane (together with corrected results by Zecca [14]) can be well approximated by the MERT fit [33]. The data of Sueoka *et al.* [34] agree with the two Trento sets above 5-6 eV but deviate downwards at lower energies; the angular resolution [34] can be again the reason for the discrepancy. In aniline, only the two Trento sets can be compared: they coincide if the energy scale correction by 0.2 eV is allowed. A bump in the total cross section just above the positronium- formation threshold (signed by an arrow in figure 4) is seen in aniline.

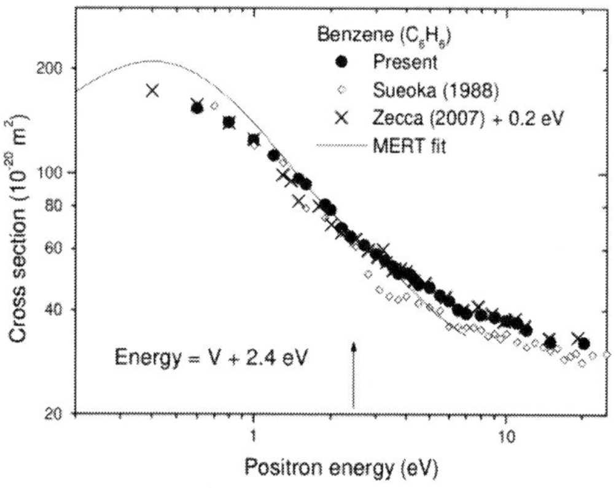

Figure 3. Benzene total cross sections - two sets from Trento are compared – Karwasz *et al.* [7] and Zecca *et al.* [14], the latter shifted by +0.2 eV. The two set coincide with the statistical error bars. The data of Suoeka [22] also coincide, up to 2.7 eV, above where a higher magnetic field was used (9 G instead of 4.5 G). MERT fit from Ref. [33].

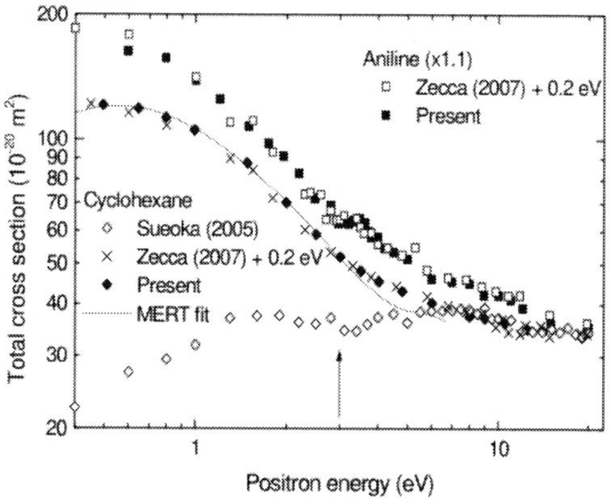

Figure 4. Cyclohexane and aniline cross sections - two experiments from Trento, Karwasz *et al.* [7], and Zecca *et al.* [14] shifted by +0.2 eV. The two set coincide with the statistical error bars. The data of Sueoka *et al.* [34] in cyclohexane are lower that the two Trento sets at low energies, probably due to poorer angular resolution. The aniline data, for clarity, were multiplied by 1.1 factor.

In benzene, Sueoka [22] measured a significant contribution to the total cross section coming from the positronium formation (about 10%) at about the same energies. If a similar contribution was the case of aniline, the bump in the total cross sections would account for the positronium formation process.

In figure 5 we come back to our reporting on resonant structures in He at 1.7 and 2.4 eV [9]. Zecca collaborated at this discovery but successively dissociated, with arguments based exclusively on post-experiment speculations, as we showed in [10]. The most recent experiment from ANU [16], performed with 530 G magnetic field has not confirmed our findings. However, it yielded some single experimental points above the average level at 2 eV and 8 eV. More experiments are needed.

A surprise comes from comparison of our helium data with the most recent, convergent close-coupling, two-centre expansion by Utamuratov *et al.* [35]. Our data agree perfectly with them, both in amplitude and the energy position, but only up to the position of resonances. The theory starts to rise visibly at 2 eV. We recall our original hypothesis [36] that the observed structures mark the opening of a new (i.e. the virtual positronium- formation) scattering channel.

Finally, let us analyze results from Trento on polar molecules, like H_2O [11] and HCOOH [13]. These molecules show a sharp rise in the cross sections in the zero-energy limit, due to their polar character. The recent theory for HCOOH by Bettega and Lima [13] give cross sections significantly higher than those reported from the experiment by Zecca *et al.* [13]. The difference is a high as 30% at 0.5 eV, see figure 3 in [13]. Several possible reasons for the discrepancy were discussed in that paper [13] but not the experimental energy-scale shift. In figure 6 we compare the Bettega and Lima's theory [13] with Zecca's experimental values [13] shifted by +0.3 eV: the two sets agree quite well.

In water the picture is particularly complicated [17]. To our knowledge, for positrons only two theories, single-centre expansion by Gianturco *et al.* [37] and Baluja *et al.* [38] in R-matrix calculations exist in the low energy range. The two theories almost coincide. More theoretical results are available for electron scattering [39, 40], but rather in disagreement, see [41]. However, recently a new experimental result [42] has been obtained for electron scattering using an absolute method and a very precise energy calibration (electrons emitted from photoionization of Ar). Zhang *et al.* [17] have calculated the integral elastic cross sections in R-matrix approximation; their results coincide with the experiment [42]. In the zero energy limit electron (and positron) scattering is dominated by the dipole interaction, so two projectiles should yield similar cross sections. Therefore, in figure 6, apart from experimental data of Zecca *et al.* [11] shifted again by +0.3 eV, we show the rotationally elastic cross sections by Zhang *et al.*: the two sets almost coincide. Resuming, the measurements on the Trento apparatus performed by Zecca can be validated, but only if the energy scale is carefully re-defined.

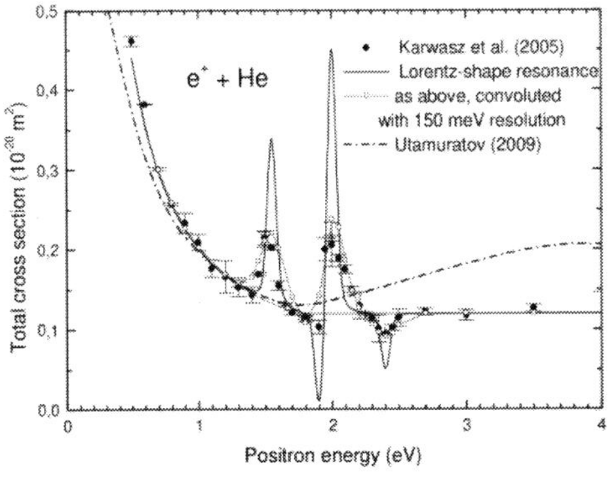

Figure 5. The virtual positronium resonance in He [9, 10, 36] compared to convergent close-coupling theory of Utamuratov *et al.* [35]. The thin lines correspond to a rigid-sphere resonance profile [36], but with the inverted resonant phase-shift ("sticky ball") and 150 meV energy resolution convolution.

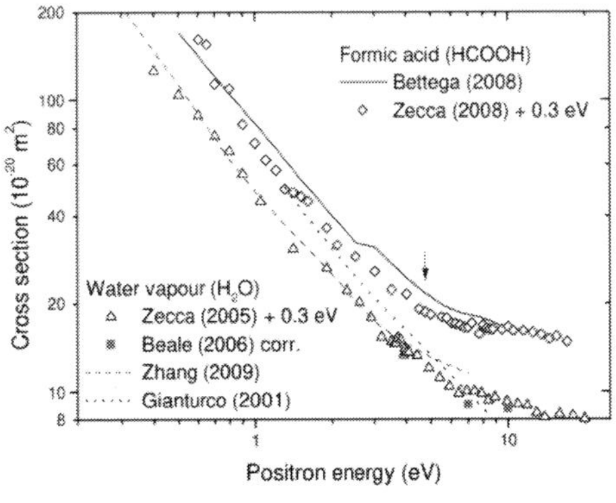

Figure 6. Positron-scattering cross sections: Zecca *et al.* for H_2O [11] and HCOOH [13], corrected by +0.3 eV; Beale *et al.* [44] as corrected by Zhang *et al.* [17]. Theory: HCOOH, Bettega and Lima [13]; H_2O, Gianturco et al. [37], Zhang *et al.* [17] electron scattering, in agreement with [42].

Acknowledgements
We thank to Dr Utumuratov, Prof. M.A.P. Lima and Prof. J. Tennyson for sending their calculations.

References

[1] Kauppila W E, Stein T S and Jesion G 1976 *Phys. Rev. Lett.* **36** 580
[2] Stein T S, Kauppila W E, Kwan C K, Parikh S P and Zhou S 1992 *Hyperfine Inter.* **73** 53
[3] Uhlmann K, Triftshäuser W, Kögel G, Sperr P, Britton D T, Zecca A, Brusa R S, Karwasz G P 1995 *Fresenius J. Analytical Chem.* **353** 594
[4] Brusa R S, Karwasz G P, Bettonte M and Zecca A1997 *Appl. Surf. Sci.* **116** 59
[5] Karwasz G P, Barozzi M, Brusa R S and Zecca A 2002 *Nucl. Instr. Meth. B* **192** 157
[6] Karwasz G P, Brusa R S, Barozzi M and Zecca A 2000 *Nucl. Instr. Meth. B* **171** 178
[7] Karwasz G P, Pliszka D, Brusa R S and Perazzolli C 2005 *Acta Phys. Pol.* **107** 666
[8] Karwasz G P, Pliszka D and Brusa R S 2006 *Nucl. Instr. Meth. B* **247** 68
[9] Karwasz G P, Pliszka D, Zecca A and Brusa R S 2005 *Nucl. Instr. Meth. B* **250** 666
[10] Karwasz G P , Brusa R S and Pliszka D 2006 *Nucl. Instr. Meth. B* **251** 520
[11] Zecca A, Sanyal D, Chakrabarti M and Brunger M J 2006 *J. Phys. B* **39** 1597
[12] Zecca A, Perazzolli C and Brunger M J 2005 *J. Phys. B* **38** 2079
[13] Zecca A, Chiari L, Sarkal A, Lima M A P, Bettega M H F, Nixon K L and Brunger M J 2008 *Phys. Rev.* **78** 042707
[14] Zecca A, Moser N, Perazzolli C, Salemi A and Brunger M J 2007 *Phys. Rev. A* **76** 022708
[15] Sullivan J P, Gilbert S J, Marler J P, Greaves R G, Buckman S J and Surko C M 2002 *Phys. Rev. A* **66** 042708
[16] Sullivan J P, Makochekanwa C, Jones A, Caradonna P and Buckman S J 2008 *J. Phys. Conf. Ser.* **133** 012004
[17] Zhang R, Faure A and Tennyson J 2009 *Physica Scripta* **80** 015301
[18] Kennerly R E 1980 *Phys. Rev. A* **21** 1876
[19] Schulz G J 1962 *Phys. Rev.* **125** 229
[20] Zubek M, Mielewska B, Channing J, King G C and Read F H 1999 *J. Phys. B* **32** 1351
[21] Sullivan J P, Gilbert S J, Buckman S J and Surko C M 2001 *J. Phys. B* **34** L467
[22] Sueoka O 1988 *J. Phys. B* **21** L631
[23] Szłuińska M, Laricchia G 2004 *Nuclear Instr. Meth. B* **221** 100
[24] Buckman S J, ICPEAC Kalamazoo invited lecture, ref. to be inserted
[25] Amarenda G, Canter K F and Schoepf D C 1996 *J. Appl. Phys.* **80** 4660
[26] Karwasz G P, Brusa R S, Piazza A and Kogel G 1995 European Community Reasearch, Project BREU-CT 90-0347, Final Technical Report, Appendix A7.1
[27] Colyert C J, Vizcaino V, Sullivan J P, Brunger M J and Buckman S J 2007 *New J. Phys.* **9** 41
[28] Dampc M, Linert I, Milosavljevic A R and Zubek M 2007 *Chem. Phys. Lett.* **443** 17
[29] Gianturco F A, Jain A and Rodriguez-Ruiz J A 1993 *Phys. Rev. A* **48** 4321
[30] McEachran R P and Stauffer A D 2009 *J. Phys. B* **42** 075202, and private information
[31] Idziaszek Z and Karwasz G P 2006 *Phys. Rev. A* **73** 064701
[32] Gribakin G and Ludlov J 2004 *Phys. Rev. A* **70** 032720 and private information, see Karwasz G P, Brusa R S, Pliszka D, Idziaszek Z, Nowakowska H 2006 *Acta Phys. Pol.* **110** 605
[33] Karwasz G P, Karbowski A, Idziaszek Z, Brusa R S 2008 *Nucl. Instr. and Meth. B* **266** 471
[34] Sueoka O, Makochekanwa C, Tanino H and Kimura M 2005 *Phys. Rev. A* **72** 042705
[35] Utamuratov R, Kadyrov A S, Fursa D V and Bray I, XV ICPEAC, Kalamazoo, https://www.icpeac2009.physics.wmich.edu/webpdffiles/Fr126.pdf, to be inserted
[36] Kawasz G P 2005 *Eur. Phys. J. D* **35** 267; Karwasz G P 2006 *Eur. Phys. J. D* **37** 153
[37] Gianturco F A, Mukherjee T and Occhigrossi A 2001 *Phys. Rev. A* **64** 032715
[38] Baluja K L, Zhang R, Franz J and Tennyson J 2007 *J. Phys. B* **40** 3515
[39] Ingr M, Polášek M, Čarsky P and Horáček J 2000 *Phys. Rev. A* **62** 032703
[40] Machado L E, Mu-Tao L, Brescansin L M, Lima M A P and McKoy V 1995 *J. Phys. B* **28** 467
[41] Karwasz G P, Zecca A and Brusa R S 2001 *Riv. Nuovo Cim.* **24** 1
[42] Čurík R, Ziesel J P, Jones N C, Field T A and Field D 2006 *Phys. Rev. Lett.* **97** 123202
[43] Humberston J W 1978 *J. Phys. B* **11** L343
[44] Beale J, Armitage S and Laricchia G 2006 *J. Phys. B* **39** 1337

Differential Ionization Studies by Positron Impact

Á. Kövér[1,2], D.J. Murtagh[1], A.I. Williams[1] and G. Laricchia[1]

[1]UCL, Department of Physics and Astronomy, Gower Street, London, WC1E 6BT, UK.
[2]Institute of Nuclear Research of the Hungarian Academy of Sciences (ATOMKI),Debrecen, Hungary

Abstract. The results of investigations of ionization of atoms and molecules by positron impact are presented, with particular focus on the electron capture to the continuum phenomenon. The application of recoil ion momentum spectroscopy in positron collisions are also discussed.

1. Introduction

During the last two decades several double and triple differential measurement of positron impact ionization have been carried out at UCL. One of the main aim of these studies has been to investigate electron capture to the continuum (ECC), a special case of ionization in which the ionized electron is strongly influenced by the positively charged projectile. This process, whose description requires higher order approximations [1] can be considered as an extrapolation across the ionization limit to highly-excited bound-states. In the case of ion impact, a sharp peak appears in the doubly-differential electron-spectrum around 0° at a velocity which is close to that of the scattered projectile. Due to the positive charge of the positron, similar final state interactions are expected. However, due to the equal masses of the electron and the positron, different collision dynamics may occur resulting in scattering for positrons to much larger angles than for heavier projectiles.

Differential investigations, especially in positron physics, are very time consuming. In the case of traditional methods, the energy of the outgoing particle is measured using electrostatic/magnetic analyzers which examine the intensity at different energies separately. Similarly, the angular distribution is scanned by rotating the analyzer around the collision region. This can lead to data collection taking weeks if not months. In order to reduce the data collection time or increase the accuracy of the measured data, new methods must be employed. By measuring the time of flight of the fragments, information on their longitudinal energy may be obtained simultaneously, but this requires a start signal. There are some examples of simultaneous angular detection by placing detectors at different angles (e.g. [2]). A recently developed system, called Cold Target Recoil Ion Momentum Spectroscopy (COLTRIMS) [3], combines the advantages of both above mentioned methods by measuring the energy and angular distribution of the recoil ion simultaneously over the full 4π-solid angle. Fig. 1 shows this principle. When the scattered projectiles are recorded in coincidence with the outgoing fragments, a kinematically complete picture can be determined about the correlated motion of atomic and molecular breakup processes. This method has been successfully used with ions, electrons and photons as projectiles (e.g. [4,5] and references therein).

In this report, results for double and triple differential cross-sections for positron impact ionization, measured using traditional methods, will be presented. This will be followed by a discussion of the requirements placed upon an experimental setup by the COLTRIM method and, finally, the progress made toward such a system will be briefly outlined.

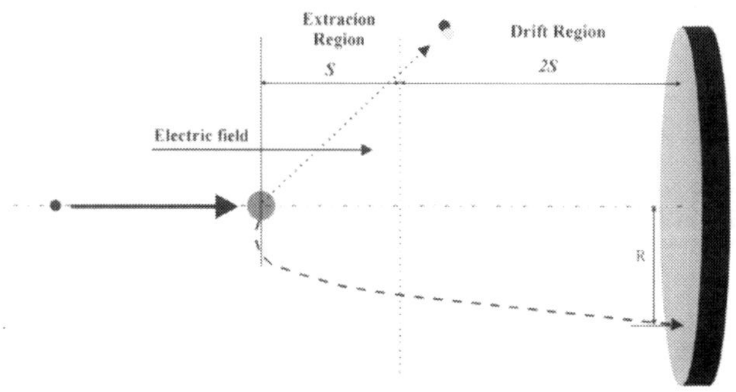

Figure 1. Basic principle of the recoil ion method.

2. Results

Initially, investigations of the doubly-differential electron energy spectra found no peak which could be associated with ECC [6,7]. Aided by theoretical investigations [8], it was surmised that the wide angle scattering of the light projectile from Ar would spread the ECC peak over a broad range of angles, in contract to the sharp cusp observed by ion impact at around 0° with respect to the beam direction. For this reason, a more sensitive (triply-differential) measurement was performed by measuring coincidences between the scattered positron and the ionized electron and a lower-Z target (H_2) was used. At 100 eV projectile energy, a small broad peak was found in the electronic spectrum, the first clear experimental demonstration of the occurrence of ECC by positron impact [9]. Theoretical calculations were soon able to reproduce these findings [10,11]. However, at 50 eV impact energy the electron spectrum was observed to be shifted to a lower energy by 1.6 eV relative to the theoretical value [12]. Further investigations showed that a similar shift but towards higher energies occurs in the corresponding energy spectrum of the scattered positrons (Fig. 2) [13]. Although not fully understood, a possible explanation of these surprising findings is that the polarization of the positron-electron pair in the positive Coulomb field of the residual target ion causes the ejected electron to be slowed down and the scattered positron to be accelerated.

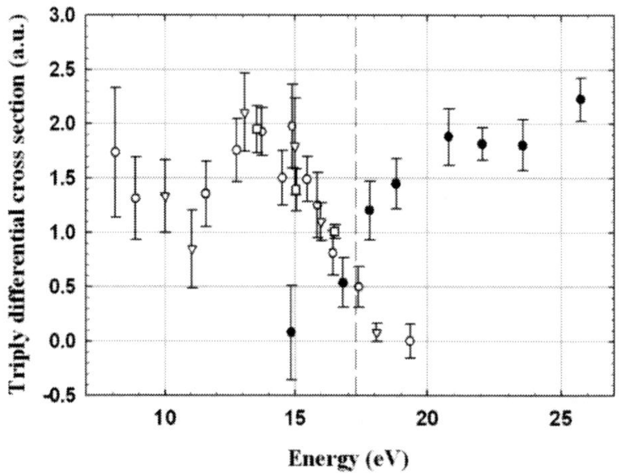

Figure 2. Experimental data of the triply differential ionization cross sections. Energy of ejected electron from 50eV e^+ collisions with, (o) H_2, (•) He and (▽) D_2. (•) indicates the energy of the scattered positrons from 50 eV e^+ - H_2 collision. The dash line shows where $v_{e+} = v_{e-}$.

Journal of Physics: Conference Series **199** (2010) 012020 doi:10.1088/1742-6596/199/1/012020

Recently, the fragmentation of H_2O has been investigated around 0° as a function of the energy loss of the e^+ projectile, at 100 and 153 eV impact energies [14]. H_2O is of special interest due to its strong dipole moment, which has been shown to cause significant forward scattering of electron projectiles [15]. The main aims of this work have been to investigate whether forward scattering enhances the probability of ECC around 0° and to measure the intensity ratios of different fragments at various projectile energies. Fig. 3 shows the results for H_2O^+ fragments at 100 eV positron impact energy. No ECC peak has been found in the double differential electron spectrum, similarly to the earlier results obtained with Ar [6], indicating that the strong dipole moment of the water does not significantly enhance the ECC process for positron projectiles. However, a small shoulder around 28 eV energy loss has been recorded at both projectile energies. This structure may be consistent with the results of high resolution electron momentum spectroscopy ([16] and references therein), identifying 27.1 eV as the onset of a weak shake-up band connected with the $2a_1$ orbital.

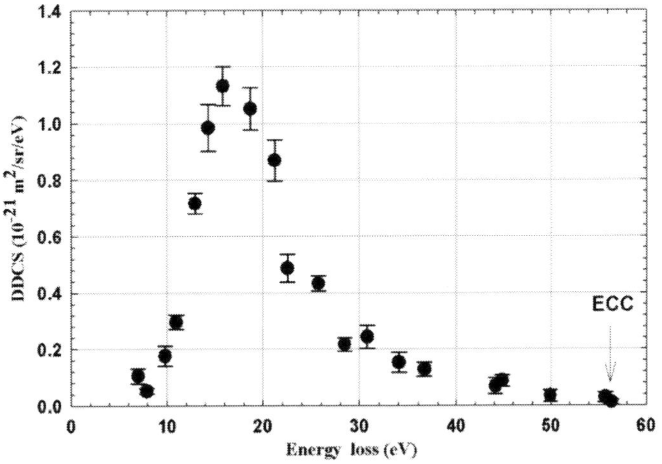

Figure 3. Energy loss spectrum of e^+ scattered projectile in coincidence with the H_2O^+ fragments. From 100 eV e^+ - H_2O collisions.

3. Discussion & Conclusion

At present, a number of groups (E.g. Williams *et al* [17], Caradonna *et al* [18]) are developing the first COLTRIMS apparatus for use with positron projectiles. The technique imposes stringent requirements upon both the projectile beam and target. To achieve the required position resolution, a parallel projectile beam with a diameter of approximately 1 mm or less is necessary. The typical diameter of an electrostatic positron beams is around 4 mm, due to the size of the moderator and the radioactive source. A high degree of accuracy is also required in the time of flight measurements. It may be possible to do this by chopping the beam (to obtain start pulses), but to obtain the required resolution the intensity of the beam would be reduced to an unacceptably-low level. For the target beam, the constraints are similar: the diameter must be small (~ 1 mm) and dense (10^{11}-10^{13} atom/cm^3); it must also have a small initial momentum distribution in order to resolve the small momentum transfer (0.3-4 a.u.) during collision. Yet application of the COLTRIMS method in positron physics has several advantages despite the technical difficulties. The 4π detection of the outgoing residual ions means that all the collision events can be detected, which greatly reduces measuring times. From the time of flight spectra, the longitudinal energy of the ions can be determined which yields information on the type of collision processes (ionization, capture to bound/excited states, etc.) whilst the ejection angle of the ions reveals the dynamics of the collision process especially when measured in coincidence with other fragments. Fig. 4 shows examples of the measurements which we plan to carry out in order to distinguish the ECC process from the Ps formation and measure the angular distributions of both processes. Here, the collision system under consideration is 100 eV e^+ - He and the extraction voltage of the ions is 3 V. The calculated time of flight can be seen on the x axes, while the y axes shows the angular deflection of the ions. From these data, the angular distribution of the Ps and the ECC process can be determined. These values are indicated on the figure.

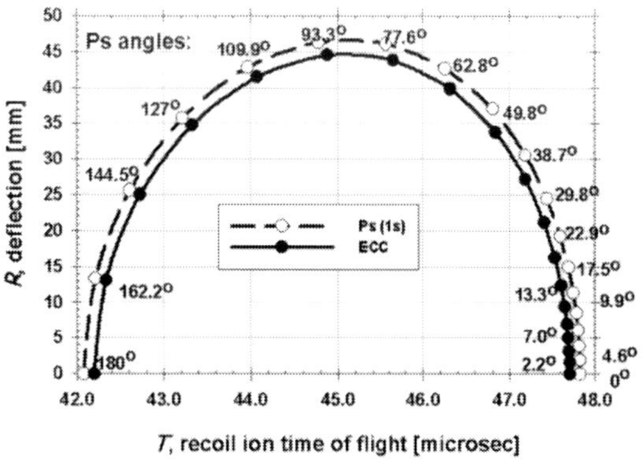

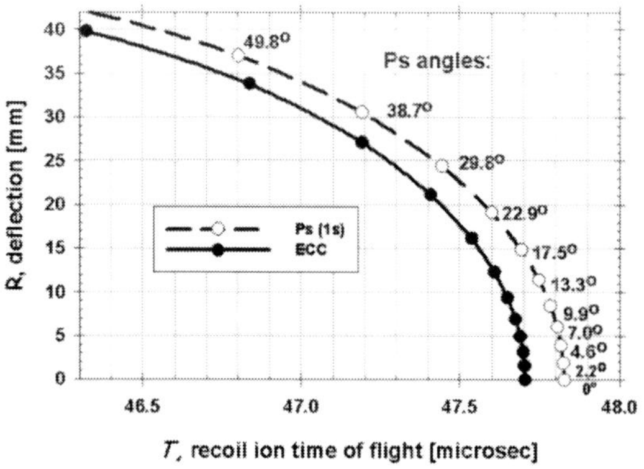

Figure 4. The calculated recoil ion time of flight and deflection from 100 eV e+ -He collisions. The numbers on the figure indicate the corresponding deflection angle of ECC/Ps fragments. The right graph shows the enlarged part of the forward scattered fragments.

To fulfill the afore mentioned requirements to undertake a COLTRIMS experiment, a new positron beam line has been developed. A Na-22 radioactive source and W mesh moderator have been combined with an electrostatic lens system to produce a 5 keV positron beam, which is tightly focused onto a 1000Å W foil remoderator. From here, the beam is transported by another electrostatic lens system to the intersection with the target beam. Included in this system is a cylindrical mirror deflector to separate the high energy beta and gamma particles from the low energy positron beam. The target is produced by a supersonic gas jet system. The interaction region is located inside an extraction spectrometer which accelerates the residual ions onto a 80 mm diameter position sensitive detector. A fuller description of this system can be found in the paper of A.I. Williams *et al* [17]. Recently a 1mm diameter e^+ beam at the interaction region has been achieved. The next step in the development of the system will be the testing of the extraction spectrometer.

Acknowledgments
This work is sponsored by Engineering and Physical Science Research Council UK (Grant no. GR/S16041/01, EP/E053521/1), Hungarian Scientific Research Foundation (NKTH-OTKA, Grant No. 67719, OTKA K73703), the European Union (HPRN-CT-2002-00179 EPIC), and COST-STSM-P9-02555.

References
[1] Brauner M., Briggs J.S., J.Phys.B:At.Mol.Opt.Phys. **24** (1991) 2227.
[2] Varga D., Kádár I., Ricz S., Végh J., Kövér Á., Sulik B., Berényi D, Nucl. Instrum. Methods A **313** (1992) 163.
[3] Ali R., Frohne V., Cocke C.L., Stockli M., Cheng S. and Raphaelian M.L.A., Phys. Rev. Lett **69** (1992) 2491.

[4] Dorner R., Mergel V., Jagutzki O., Spielberger L., Ullrich J.,. Moshammer R. and Schmidt-Bocking H., Physics Reports **330** (2000) 95.
[5] Ullrich J. , Moshammer R., Dorn A, Dörner R, H Schmidt L Ph. and Schmidt-Böcking H., Rep.Prog. Phys. **66** (2003) 1463.
[6] Kövér Á, Laricchia G and Charlton M., J.Phys.B:At.Mol.Opt.Phys. **26** (1993) L575.
[7] Kövér Á, Laricchia G. and Charlton M., J.Phys.B:At.Mol.Opt.Phys. **27** (1994) 2409.
[8] Sparrow R A, and Olson R E J.Phys.B:At.Mol.Opt.Phys **27** (1994) 2647.
[9] Kövér Á and Laricchia G., Phys. Rev. Lett. **80** (1998) 5309.
[10] Berakdar J., Phys. Rev. Lett. 81 (1998) 1393.
[11] Fiol J., Rodriguez V. D., and Barrachina R. O., J. Phys. B **34** (2001) 933.
[12] Kövér Á, Paludan K., and Laricchia G., J.Phys.B:At.Mol.Opt.Phys **34** (2001) L219.
[13] Arcidiacono C., Kövér Á., Laricchia G., Phys. Rev. Lett. **95** (2005) 223202.
[14] Arcidiacono C., Beale J., Pešic Z. D., Kövér Á. and Laricchia G., J.Phys.B:At.Mol.Opt.Phys. **42** (2009) 065205.
[15] Baluja K L, Zhang R, Franz J and Tennyson J, J. Phys. B: At.Mol. Opt. Phys. **40** (2007) 3515.
[16] Ning C G, Hajgato B, Huang Y R, Zhang S F, Liu K, Luo Z H, Knippenberg S, Deng J K and Delueze M S., Chem. Phys **343** (2008)19.
[17] Williams A., Kövér Á., Murtagh D. and Laricchia G., J. Phys: Conference Series, this proceedings
[18] Caradonna P., Jones A., Makochekanwa C., Slaughter D., Sullivan J., Mueller D., Bellm S., Lohmann B., Buckman S., J. Phys: Conference Series, this proceedings.

Journal of Physics: Conference Series **199** (2010) 012021 doi:10.1088/1742-6596/199/1/012021

Convergence study of the close-coupling approach to positron-helium collisions

R. Utamuratov, A. S. Kadyrov, D. Fursa and I. Bray
ARC Centre for Antimatter-Matter Studies, Curtin University of Technology, Perth, Australia
E-mail: r.utamuratov@curtin.edu.au

Abstract. Positron-helium collisions have been studied using a two-center close-coupling method. Convergent, pseudoresonance-free cross sections have been obtained by using complete expansions on both the helium and positronium centers. The low energy elastic cross sections, phase-shifts and total cross sections are compared with experimental data and results of other theories.

1. Introduction

The convergent close-coupling (CCC) method with a two center expansion has very successfully described positron-hydrogen collisions [1]. It has been shown that a two-center expansion with sufficient number of states from both the atom and positronium (Ps) centers yields smooth and convergent results [2].

Although the CCC method using only a single center expansion was successful in describing elastic, total and total break-up positron-helium cross sections at some energies [3, 4], it was unable to get convergent results in the Ore gap, or estimate the Ps-formation cross section. Two-center methods using only eigenstates for at least one of the centers can, in general, give good estimates. However, they often produce pseudoresonances in the cross sections. For example, close-coupling results of Hewitt *et al* [5] and Campbell *et al* [6] have suffered from pseudoresonances below 30 eV incident energies.

Studies of positron-hydrogen collisions using the two-center CCC approach have shown that pseudoresonances will be eliminated if full pseudostate expansions are used on both centers. Consequently, we expect that complete two-center expansions will be required to get pseudoresonance-free cross sections for calculating positron-helium collisions as well.

The positronium (Ps) formation channels have now been added to the CCC method for positron-helium scattering. Convergent results have been obtained for elastic scattering phase-shifts, ground and excited state Ps-formation and He excitation cross sections. We present the results for the elastic cross sections and low energy phase-shifts in order to make comparison with the benchmark results of the variational calculations of Van Reeth and Humberston [7]. Cross sections for Ps-formation, helium excitation and ionization processes will be presented elsewhere.

2. Method

The basic idea of the two-center convergent close-coupling approach is to expand the total scattering wavefunction in terms of the full set of target and positronium states. Using the expanded e^+-He total wavefunction the Schrödinger equation is reduced to a set of momentum-space coupled integral equations for the transition amplitudes, which are solved in partial-wave reduced form.

The helium and positronium states are obtained by independently diagonalizing their Hamiltonians using Laguerre bases, which will ensure square integrability of the resulting wavefunctions. By increasing the basis sizes, the negative energy states converge towards the true eigenstates, while the positive energy pseudostates yield an increasingly dense discretization of the continuum.

Convergence in the cross sections is obtained by increasing the basis sizes of both the He and Ps centers. Ps-formation processes are calculated explicitly as transitions to negative-energy Ps-states. The ionization is also calculated simply as electron capture into positive-energy pseudostates of Ps and He.

We made the following two approximations in order to simplify the calculations. Firstly, the frozen-core treatment of helium, which assumes that one of the electrons is always in the $He^+(1s)$ state [8]. Such an approximation gives a reasonably accurate ground state of He with ionization energy of 23.74 eV instead of the actual 24.58 eV, and more accurate excited states. Secondly, we neglected the electron exchange between positronium and He^+ ion. This approach is supported by Van Reeth and Humberston [9], who stated that the He^+ ion behaves similarly to a proton because of tightly bound electron in the ion and also mean kinetic energies of the two electrons are different by a factor of eight, which makes electrons effectively distinguishable.

3. Results

In our present method we can include as many eigen- or pseudo-states from both centers as computing resources allow. In this work we present results of three kinds of calculations. The first one, denoted by CC($\overline{24}$,3), is with 24 (9s,8p,7d) helium pseudostates and three (1s,2s,2p) Ps eigenstates, is similar to the calculation of Campbell *et al* [6]. The second and third ones have 27 (10s,9p,8d) and 33 (12s,11p,10d) pseudostates for both centers, denoted as CCC($\overline{27}$,$\overline{27}$) and CCC($\overline{33}$,$\overline{33}$), respectively. These two demonstrate that results the level of convergence. The CCC($\overline{33}$,$\overline{33}$) calculations are more time consuming than the CCC($\overline{27}$,$\overline{27}$) ones and therefore done at fewer energy points.

Figs. 1 and 2 presents phase-shifts below Ps-formation threshold. Our results compare well with the variational calculations [7], which are believed to be highly accurate and benchmarks. Convergence of the CCC calculations is excellent, and so by comparison with the multi-configuration single-center calculations [3], we can conclude that variance from the variational results is due to the present frozen-core treatment of helium.

Fig. 3 presents elastic cross sections below the Ps-formation threshold. Our results agree well with the experimental data and the results of variational calculations [7], except at the very low energies, due to the frozen-core approximation. The apparent good agreement of our CC($\overline{24}$,3) calculations at the lower energies with experiment is coincidental as convergence has not yet been achieved.

We note that very similar results were obtained by one center CCC calculations using very high orbitals ($l_{max} = 8$) [3]. At low and medium incident energies it is very important to properly consider strong electron-positron correlations. Therefore in one-center expansions it was required to use very high angular momenta to get convergent results. In the two center method positron-electron correlation is explicitly included as Ps-formation channels, therefore there was no need to use high angular momenta states. Although not shown, we have checked (at a few partial wave and energy points) if it is the case by including f-states, and it has insignificant effect on the results. This was seen in positron-hydrogen case as well.

Fig. 4 displays total cross section from Ps-threshold up to energies where effect of Ps-formation diminishes. CCC results agree well with the experimental data [12, 11] up to 30 eV. At the

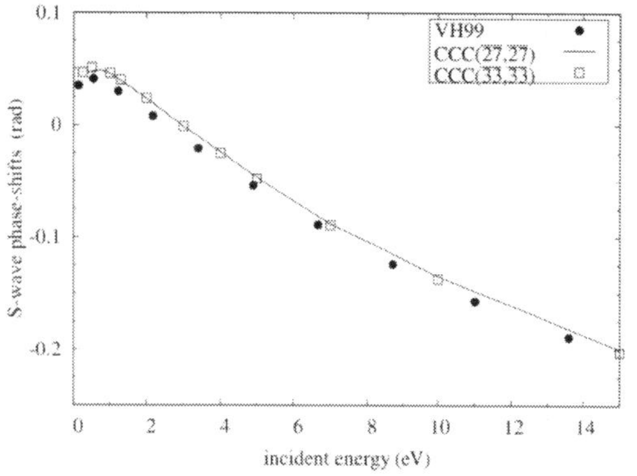

Figure 1. S-wave phase-shifts. The variational calculation of Van Reeth and Humberston [7] is denoted by VH99. The present calculations are described in the text.

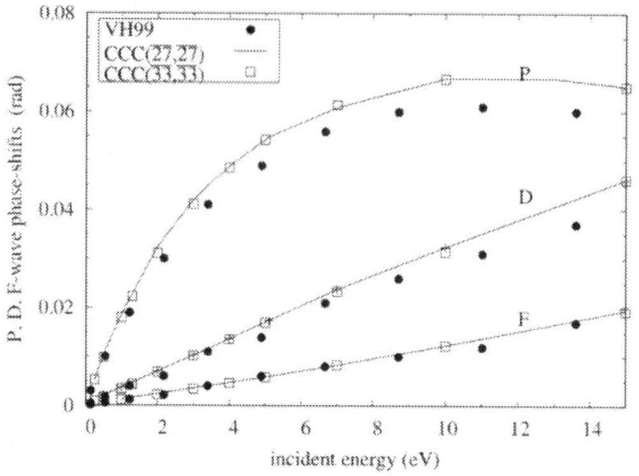

Figure 2. The same as Fig.1, but for P-, D-, and F-wave phase-shifts.

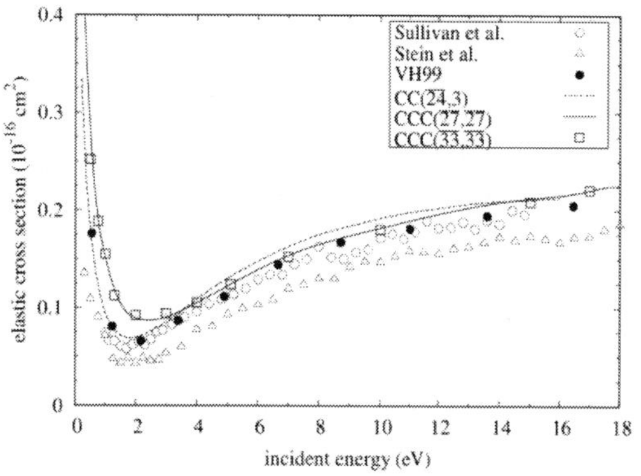

Figure 3. e^+-He elastic scattering cross section. Experimental data are due to [10] and [11]. The calculations are described in the text.

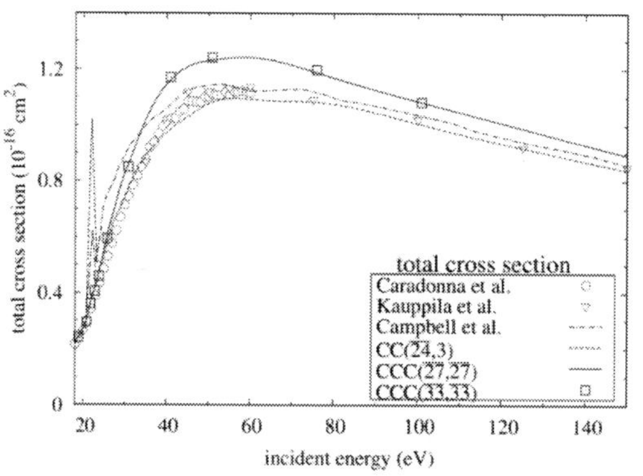

Figure 4. Total e^+-He scattering cross sections. Experimental data are due to [12] and [13]. The calculations are described in the text.

higher energies the CCC total cross section overestimates the experimental data. This is due to primarily the frozen-core treatment of He. The better agreement of CC($\overline{24}$,3) results and experimental data is coincidental as these results are not yet convergent. One remarkable feature of the CCC results is that they are smooth over the whole energy range, while our CC($\overline{24}$,3) and the results of Campbell *et al* [6] have resonance structure around 22 eV.

4. Conclusion

The two center CCC method has been applied to positron-helium scattering. Convergent cross sections were obtained by using an equal number of pseudostates for both helium and Ps centers. Low energy elastic cross sections and phase-shifts have been obtained by using only s-, p-, and d-orbitals, and the results agree well with experiments and benchmark calculations of Van Reeth and Humberston [7].

The present total cross sections are slightly higher than the experimental ones at incident energies above 30 eV. We believe that this overestimation is due to the frozen-core treatment of helium. We expect that relaxation of the frozen-core approximation will lead to a slight reduction of the corresponding cross sections, as was found in the single-center calculations of Wu *et al* [4], and so will result in even better agreement with experiment.

The work was supported by the Australian Research Council. We are grateful for access to the Australian National Computing Infrastructure Facility and its Western Australian node iVEC.

References

[1] Kadyrov A S and Bray I 2002 *Phys. Rev.* A **66** 012710
[2] Kadyrov A S and Bray I 2000 *Nucl. Instr. and Meth.* B **171** 119–125
[3] Wu H, Bray I, Fursa D V and Stelbovics A T 2004 *J. Phys.* B **37** L1–L6
[4] Wu H, Bray I, Fursa D V and Stelbovics A T 2004 *J. Phys.* B **37** 1165–72
[5] Hewitt R N, Noble C J and Bransden B H 1992 *J. Phys.* B **25** 557
[6] Campbell C, McAlinden M, Kernoghan A and Walters H 1998 *Nucl. Instr. and Meth.* B **143** 41
[7] Van Reeth P and Humberston J W 1999 *J. Phys.* B **32** 3651–67
[8] Fursa D V and Bray I 1995 *Phys. Rev.* A **52** 1279–98
[9] Van Reeth P and Humberston J W 1998 *J. Phys.* B **31** L621
[10] Stein T S, Kauppila W E, Pol V, Smart J H and Jesion G 1978 *Phys. Rev.* A **17** 1600
[11] Sullivan J P, Makochekanwa C, Jones A, Caradonna P and Buckman S J 2008 *J. Phys.* B **41** 081001
[12] Kauppila W E, Stein T S, Smart J H, Dababneh M S, Ho Y K, Downing J P and Pol V 1981 *Phys. Rev.* A **24** 725
[13] Caradonna P, Sullivan J P, Jones A, Makochekanwa C and Buckman S J 2009 *submitted for publication*

Journal of Physics: Conference Series **199** (2010) 012022 doi:10.1088/1742-6596/199/1/012022

Classical Trajectory Monte Carlo Calculation of the Fully Differential Cross Section for Ionization of H_2 by Positron Impact

R. O. Barrachina and J. Fiol

Centro Atómico Bariloche and Instituto Balseiro (Comisión Nacional de Energía Atómica and Universidad Nacional de Cuyo)
8400 S. C. de Bariloche, Río Negro, Argentina

E-mail: barra@cab.cnea.gov.ar, fiol@cab.cnea.gov.ar

Abstract. The possibility of performing kinematically complete measurements of ionization collisions by positron impact brings about the promise of encountering new unexpected phenomena, which otherwise would be impossible to foresee. In this communication, we employ the classical trajectory Monte-Carlo method to explore H_2 ionization fully differential cross section (FDCS), depending on the relative momentum of the electron-positron continuum dipole and the deflection angle of its center of mass. We find a strong structure in the FDCS at non-zero deflection angles. Finally, we discuss some of these new effects, and the possibility of observing them in actual experiments.

1. Introduction

In the early sixties experimental techniques reached a stage where it was possible to measure double differential cross sections (DDCS) on the energy and angle of the electrons ejected in ion-atom collisions. This technical breakthrough opened the way to the discovery of new unforeseen phenomena. The most conspicuous was a pronounced peak that occurred when the electron-projectile relative momentum $\boldsymbol{k}$ approached zero [1, 2, 3]. Many years later this cusp, first thought to be due to spurious electrons [1], was identified and explained in terms of an electron capture to the continuum (ECC) mechanism [4]. In positron impact collisions, this peak is spread out in the DDCS by the deflection of the much lighter projectile [5, 6], and only becomes visible if the projectile scattering angle is also measured, as demonstrated by Kövér and Laricchia in 1998. Further experiments [7, 8] disclosed another interesting effect: The ECC cusp was not centered at $k = 0$. While $k = 0$ corresponds to electron and positron matching energies, it was experimentally observed that the cusp maximum is shifted to lower values of the electron energy and higher values of the positron energy.

2. Classical Trajectory Monte Carlo (CTMC) Simulations

The natural conclusion of these few examples is that by increasing the dimension of the cross section, new effects are discovered. Thus, the possibility of performing kinematically complete measurements of ionization collisions by positron impact brings about the possibility of encountering new phenomena, which otherwise would be impossible to foresee. With this premise in mind, let us explore the Fully Differential Cross Section (FDCS) for ionization of molecular hydrogen by positron impact. To this end we employ "Classical Trajectory Monte Carlo" (CTMC) simulations with more than 2×10^8 trajectories [9]. We study the ionization of H_2 molecules by the impact of 50 keV positrons using a simple three-body model [9]. The obtained cross sections correspond to averaged values over all possible orientations of the molecule. By rotational symmetry and momentum and energy conservation, this FDCS is fully characterized by four out of the nine variables associated to the three-body final state [10]. A convenient choice to study the ECC cusp features is the reduced electron-positron momentum $\boldsymbol{k} = (\boldsymbol{k}_e - \boldsymbol{k}_P)/2$ and the direction of the momentum $\boldsymbol{K}$ of the positron-electron center-of-mass respect to the residual target ion.

Figure 1 shows the cartesian projections of the cross section in $\boldsymbol{k}$-space. The three upper maps correspond to the center-of-mass (CM) of the electron-positron system restricted to moving within 5 degrees of the direction of the initial positron velocity $\boldsymbol{v}$, while in the lower maps the CM has been deflected in 45 ± 5 degrees.

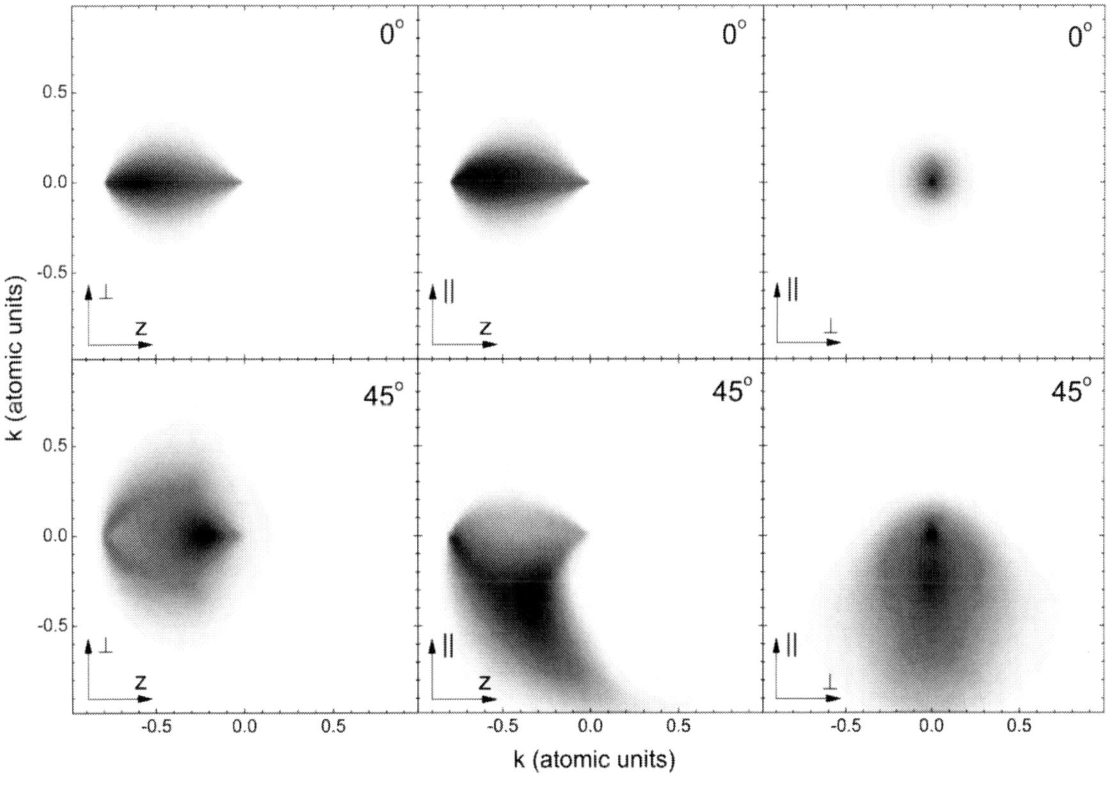

Figure 1. Cartesian projections of the distribution of the electron - positron relative momentum $\boldsymbol{k}$ for the ionization of H_2 molecules by 50 keV positron impact. The orthogonal components of $\boldsymbol{k}$ are defined as follows: k_z is parallel to the momentum $\boldsymbol{K}$ of the electron-positron center-of-mass with respect to the residual target ion; $k_{||}$ and $k_{\perp}$ are parallel and perpendicular to the plane formed by $\boldsymbol{K}$ and the initial velocity $\boldsymbol{v}$, respectively. The figures correspond to angles of 0° and 45° between $\boldsymbol{K}$ and $\boldsymbol{v}$ as indicated, within an uncertainty of 5°.

The momentum distribution in figure 1 is peaked at $k = 0$ and $\boldsymbol{k} \approx -0.8\,\hat{\boldsymbol{z}}$ a.u., but blurred to some extent by the resolution employed to visualize the CTMC simulation. The first cusp corresponds to the aforementioned ECC effect, while the second can be related to the excitation of the electron to a low-lying continuum state of the target. Another conspicuous feature of this momentum distribution is a strong orientation of the electron-projectile pair away from the forward direction ($k_Z > 0$), an effect that can be ascribed to the Coulomb interaction with the receding ion. It was recently proposed that this effect might explain why, while centered at $k = 0$ as it was just explained, the ECC cusp exhibits an apparent deviation towards lower electron energies when it is convoluted with the experimental acceptance [11]. In figure 2 we show this shift of the ECC cusp at 0° for the FDCS as a function of the electron energy and angle and the positron angle. The cross sections for larger final angles of the electron and the positron in a coplanar geometry show increasingly larger displacements. While the cusp should be located at exactly the same energy for all angles by energy and momentum conservation, this increasing shift represents a characteristic fingerprint of the strong orientation of the electron-projectile continuum dipole previously described.

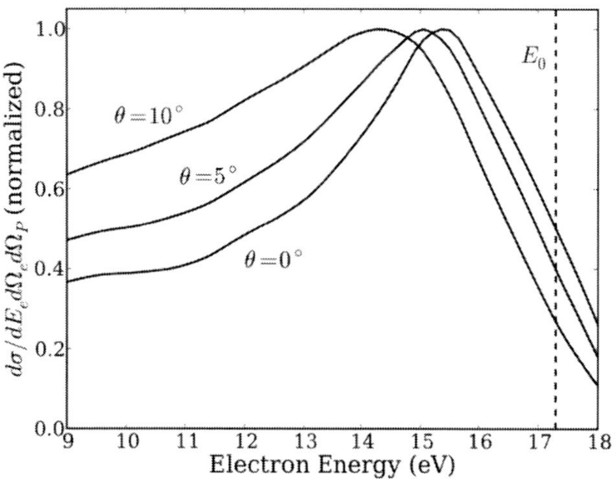

Figure 2. Normalized cross sections for the ionization of H_2 molecules by 50 keV positron impact. The electron is ejected in the plane determined by the initial and final positron momenta in the Lab frame. The electron and the positron are detected in the same direction at an angle θ with respect to the positron incident velocity direction. The plane is determined in all cases with an uncertainty of 30°. E_0 is the energy corresponding to the kinematical position of the cusp, obtainable only for infinitesimal resolution. The cusp should be located at exactly this energy for all emerging angles.

These are only some of the structures observed in figure 1. While at 0° the cross section is relatively simple due to the rotational symmetry of the collision process, it becomes very complex at non-zero deflection angles, up to the point that any subjacent structure is difficult to apprehend. In cases like this, rotating the point of view might help to get a clearer visualization, as shown in the video accompanying the online version.

Structures like the one described above are not only difficult to identify and visualize in CTMC and Quantum Mechanical calculations, but also very hard to measure. However, they might still leave observable fingerprints in lower-dimension cross sections. For instance, the DDCS in the recoil ion momentum exhibits a threshold structure that can be related to the ECC cusp at k=0, as shown in figure 3. This threshold has even an effect on the cross section singly differential in the parallel component [12], as it can be seen in the upper panel of figure 3, a structure than becomes sharper whenever the transversal component is restricted to a narrow region in momentum space, (see figure 4). These results might represent good opening experiments for future positron cold-target recoil-ion momentum spectroscopy (ColTRIMS) setup.

3. Conclusions

We have calculated the FDCS for the ionization collisions by positron impact by means of CTMC simulations. We have discussed some effects observed in this very complex cross section. They display some structures, some very well known, such as the ECC cusp, some novel, such as the

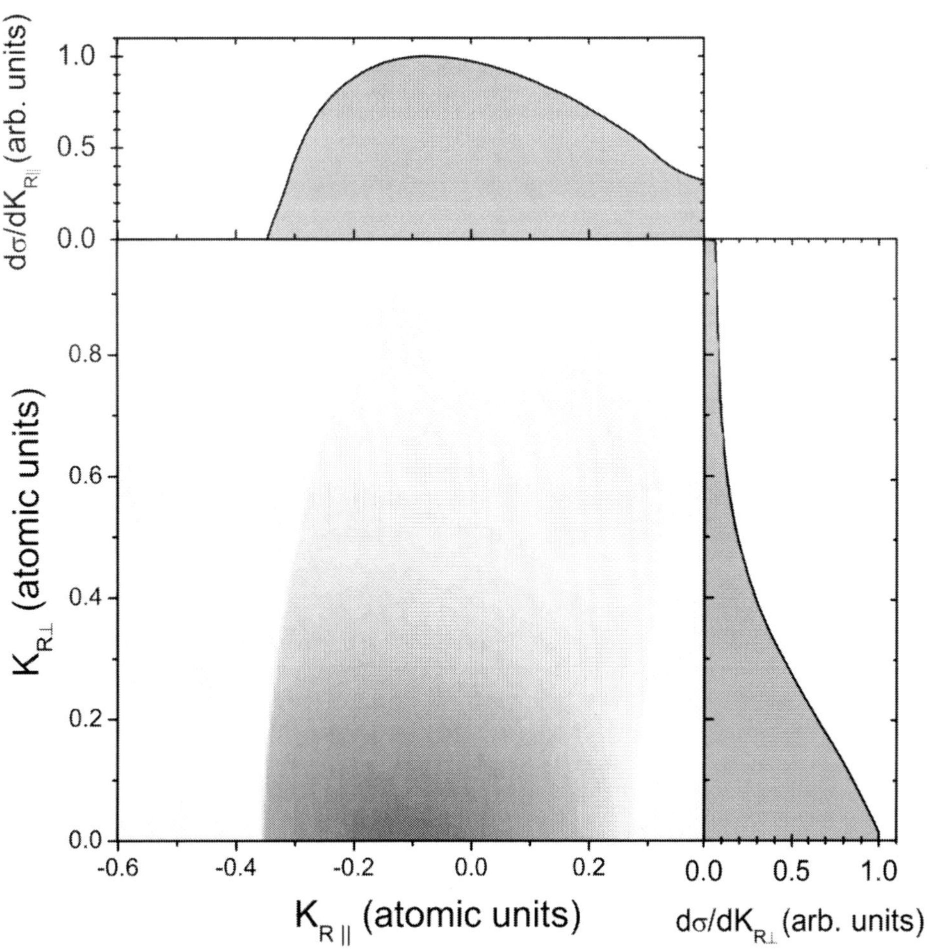

Figure 3. The gray scale map displays the DDCS for the ionization of H_2 molecules by 50 keV positron impact as a function of the components of the recoil-ion momentum $\boldsymbol{K}_R$ parallel (||) and perpendicular ($\perp$) to the initial projectile velocity $\boldsymbol{v}$. The upper and right side graphs show the cross sections single differential in $K_{R||}$ and $K_{R\perp}$, respectively.

orientation of the electron-positron system. It is noteworthy that this orientation is reproduced in simulations for ionization at higher incident energies, though the effect is less pronounced. Also, calculations on ionization of helium by impact of 50 eV positrons show a similar effect. Finally we discussed how these effects might affect lower dimension cross sections, making it feasible to observe them in actual experiments.

Acknowledgments
This work was partially supported by the Agencia Nacional de Promoción Científica y Tecnológica (Grant PICT 03-20548), the Consejo Nacional de Investigaciones Científicas y Técnicas (Grant PIP 112-200801-01269) and Universidad Nacional de Cuyo (Grant 06/C229), Argentina. The authors are also members of the Consejo National de Investigaciones Científicas y Técnicas (CONICET), Argentina.

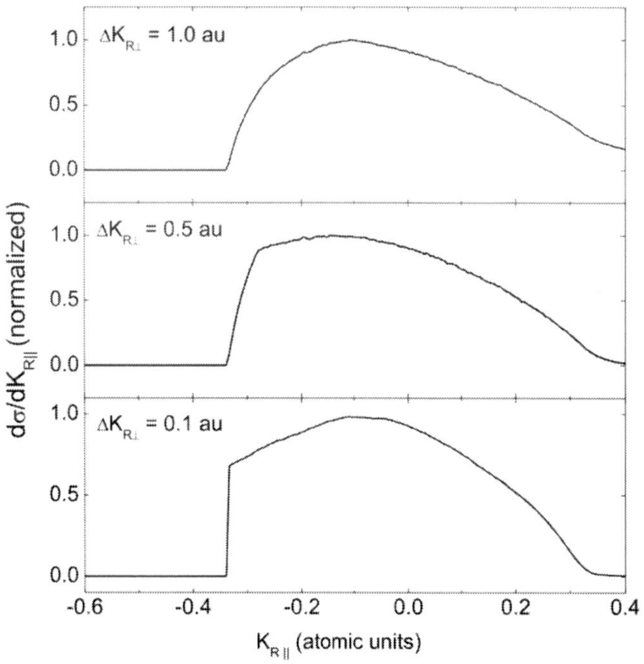

Figure 4. Normalized cross section for the ionization of H_2 molecules by 50 keV positron impact, single differential in the recoil ion momentum component parallel to the initial velocity $\boldsymbol{v}$. The threshold becomes sharper for decreasing acceptances $\Delta K_{R\perp}$ of the transversal momentum component. In the limit of perfect resolution the cross section presents a discontinuity in the threshold.

References

[1] Rudd M E and Jorgensen T 1963 *Phys. Rev.* **131** 666–675
[2] Rudd M E, Sautter C A and Bailey C L 1966 *Phys. Rev.* **151** 20–27
[3] Crooks G B and Rudd M E 1970 *Phys. Rev. Lett.s* **25** 1599–1601
[4] Macek J 1970 *Phys. Rev. A* **1** 235–241
[5] Sparrow R A and Olson R E 1994 *J. Phys. B* **27** 2647–2655
[6] Fiol J, Rodríguez V D and Barrachina R O 2001 *J. Phys. B* **34** 933–944
[7] Kövér Á, Paludan K and Laricchia G 2001 *J. Phys. B* **34** L219–L222
[8] Arcidiacono C, Kövér Á and Laricchia G 2005 *Phys. Rev. Lett.s* **95** 223202
[9] Fiol J and Olson R E 2002 *J. Phys. B* **35** 1173–1184
[10] Della Picca R, Fiol J, Barrachina R O and Rodríguez V D 2006 *Nuc. Inst. Meth. B* **247** 52–57
[11] Fiol J, Macri P and Barrachina R O 2009 *Nuc. Inst. Meth. B* **267** 211–214
[12] Barrachina R O, Fiol J and Macri P 2008 *Nuc. Inst. Meth. B* **266** 402–406

Low energy elastic scattering of positrons by argon atoms

Harsh Mohan[1], Arvind Kumar Jain[1] and Sunita Sharma[2]

[1]Physics Department, M.L.N. College, Yamuna Nagar 135 001, Haryana, India
[2]Chemistry Department, M.L.N. College, Yamuna Nagar 135 001, Haryana, India

e-mail: mohan_harsh@yahoo.com

Abstract. Positron scattering is complementary to electron scattering due to the difference in polarity of charge. This makes positrons an alternative tool to study atomic and molecular structure. In this paper, we report our calculations of differential and total cross sections for the elastic scattering of positrons with argon atoms in the energy range from 1 – 10 eV. A spherical optical potential approach is employed. The static potential is included exactly at the Hartree-Fock level, the short-range correlation and long-range polarization effects are target density functional and are free from any adjustable parameter. Our results are in fair agreement between experimental measurements and other theories.

1. Introduction

Study of electrons and positrons with atoms, ions and molecules can provide useful information for understanding of their structures, reaction mechanism and a lot of actual and prospective applications in the field of technology, environment and the biomedical sciences. The last decade is witness to a tremendous revival of interest in the general area of scattering processes involving positrons as projectiles with atoms and simple molecules in the gaseous phase as targets [1]. A variety of possible outcomes from the interaction of positrons obviously parallels that for electron projectiles. But in case of low energy positrons interactions, there is scarcity of experimental measurements, which is expected if the difficulties linked with weak yield of positron from radioactive sources are considered. However recently, there has been much attention in the experimental aspects of the scattering of positrons. This is because the dynamical processes involving energy losses and quenching of positron beam passing through either matter or molecular gases have gained importance in diverse fields ranging from the breakdown in the gaseous media to plasma chemistry. Consequently, an updated description of both elastic and inelastic scattering of positrons by atoms has always been in demand. More precisely, the accurate cross sections of elastic scattering events are highly desirable in Monte-Carlo simulations.

Initially, the collision cross sections observations with positron scattering were included in the 'first generation experiments' [2]. It was then observed that the scattering of positrons by only the rare-gas atoms exhibits the Ramsauer-Townsend (RT) minima. Thus argon may play an important role in showing this feature. Several groups [3-8] have calculated elastic scattering cross sections of electron/positron from the noble gases. In continuation of this, McEachran and Stauffer [9] have calculated differential cross sections for argon atoms using complex *ab initio* optical potential. This motivates us to do current work in low energy regime. This work is in continuation of our previous calculations [10]. In this paper, we demonstrate differential cross sections (DCS) and total cross sections (σ_t) for the elastic scattering of positrons by argon atoms in the energy range from 1 – 10 eV using spherical optical potential (SOP) model approach [11].

2. Methodology

In the usual potential scattering problem, the following Schrodinger-equation is solved for the scattered particle at energy k^2 i.e.

$$\left[\frac{d^2}{dr^2} + k^2 - \frac{\ell(\ell+1)}{r^2} - V_R(r)\right] f_\ell(kr) = 0 \tag{1}$$

where $V_R(r)$ is the real potential for the e^+ - Ar system written as

$$V_R(r) = V_{st}(r) + V_p(r) \tag{2}$$

The static potential is the average over the ground state atomic charge distribution of the electrostatic interaction of the positron and atom. The static potential is calculated accurately from

$$V_{st}(\boldsymbol{r}) = \frac{Z}{r} - \int \frac{\rho_0(r')d^3r'}{|\mathbf{r} - \mathbf{r}'|} \tag{3}$$

where $\rho_o(r)$ is the target unperturbed charge density.

The polarization effect arises from the distortion of the target electronic wave due to the incident charged particle. For the elastic scattering, the polarization-correlation contribution accounts for all virtual transitions to the excited states. We have used a parameter-free polarization potential (V_p), which is based on the correlation energy of the target atom [12]. It has two components, the short-range [$V_{SR}(r)$] and the long-range [$V_{LR}(r)$] parts, and is given by

$$V_p(r) = \begin{cases} V_{SR}(r), & r < r_c \\ V_{LR}(r), & r \geq r_c \end{cases} \tag{4}$$

Here, r_c is the point where two forms cross each other for the first time. The crossing point for argon atoms occurs at 2.90 a.u. The short-range form for the positron scattering with atomic systems is given by

$$V_{SR}(r) = \begin{cases} 0.0622\, ln\, r_s + 0.018\, r_s\, ln\, r_s - 0.02\, r_s - 0.096, & r_s \leq 0.7 \\ 0.03796\, ln\, r_s - 0.1231, & 0.7 \leq r_s \leq 10.0 \\ -0.876\, r_s^{-1} + 2.65\, r_s^{-3/2} - 2.8\, r_s^{-2} - 0.8\, r_s^{-5/2}, & 10.0 \leq r_s \end{cases} \tag{5}$$

The long-range form of the polarization is taken as the following well known asymptotic form,

$$V_{LR}(r) = -\frac{\alpha_d}{2\,r^4}, \quad r > r_c \tag{6}$$

where α_d is the dipole polarizability of the target atom. For the present calculations, its value is taken as 11.08 a.u. for the argon atom [13].

Both potential terms are function of electronic density of the target. The static potential (V_{st}) and the charge density (ρ_o) are obtained using non-relativistic Slater-type orbital of Roothan–Hartree Fock wave functions as given by McLean and McLean [14].

We have employed the variable - phase approach (VPA) as proposed by Calagero [15] to find the solution of the equation (1). The corresponding quantities i.e. DCS and σ_t are then easily obtained from phase shifts at each energy by this method. All our phase shifts are converged with respect to the number of partial waves up to a value of 0.00001 radians.

3. Results and discussion

Figure 1 shows our elastic DCS for positron scattering with argon atoms at 5 eV. The results are compared with the theoretical calculations of McEachran and Stauffer [9] as well as elastic DCS of Nahar and Wadehra [16] along with experimental data of Kauppila *et al* [17]. It has been found that the present results are better than the calculations of Nahar and Wadehra [16]. These are also in agreement with the calculations of McEachran and Stauffer [9] as well as experimental measurements of Kauppila *et al* [17] reproducing a single broad minima.

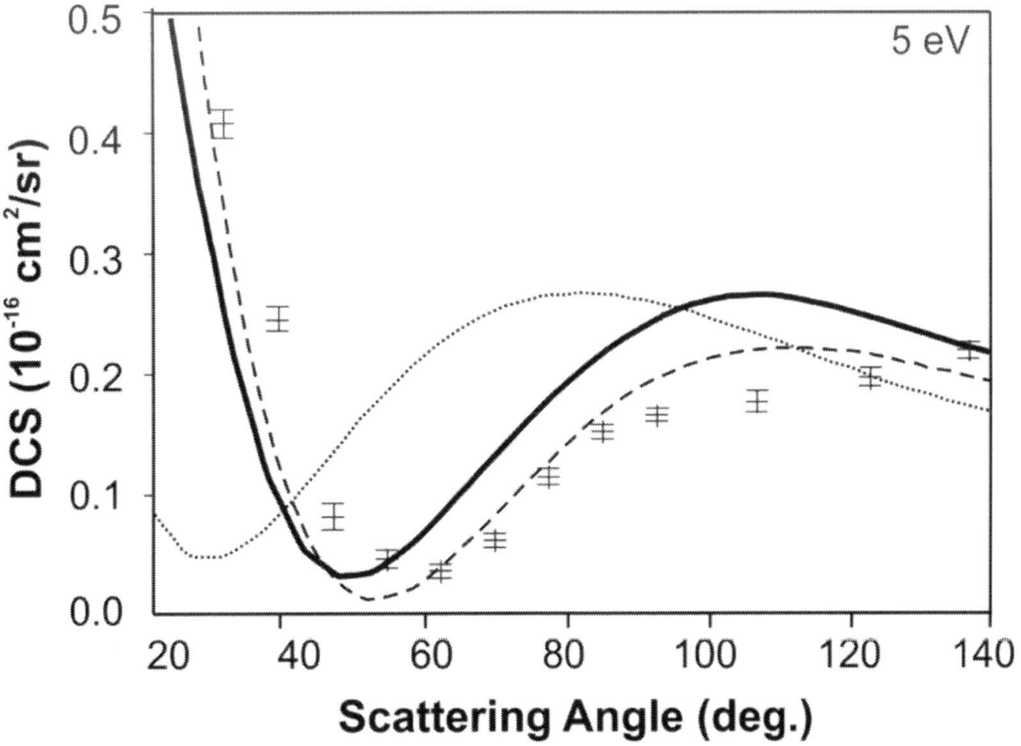

Figure 1. Differential cross sections for e^+ - Ar scattering at 5 eV. Calculations: ____, present results; - - -, McEachran and Stauffer [9]. ....., Nahar and Wadehra [16]. Experiment: ⫫, Kauppila *et al* [17].

In figures 2, we have compared our DCS results with the recent calculations of McEachran and Stauffer [9] along with experimental data of Kauppila *et al* [17] at 8.7 eV. It is noticed that there exist a variation in the magnitude of the DCS between both theoretical calculations and experimental measurements at small scattering angles. This is reflected as a shift in the minima towards the forward direction i.e. upto $\approx 40^o$. But the difference narrows down for scattering angle $> 40^o$ extending for large angles (backward direction). Although this is a disagreement between theoretical calculations and experimental data but on the other hand retrace prevailing trends.

Total cross sections (σ_t) are also evaluated from the DCS's. In figure 3, the σ_t cross sections are presented and compared with the latest experimental measurements of Coleman *et al* [18] in the energy range from 1 – 10 eV. It has been observed that our theoretical results are in good agreement with the experimental data below threshold energy. The disagreement may be due to the inelastic channels which play important role beyond threshold energy.

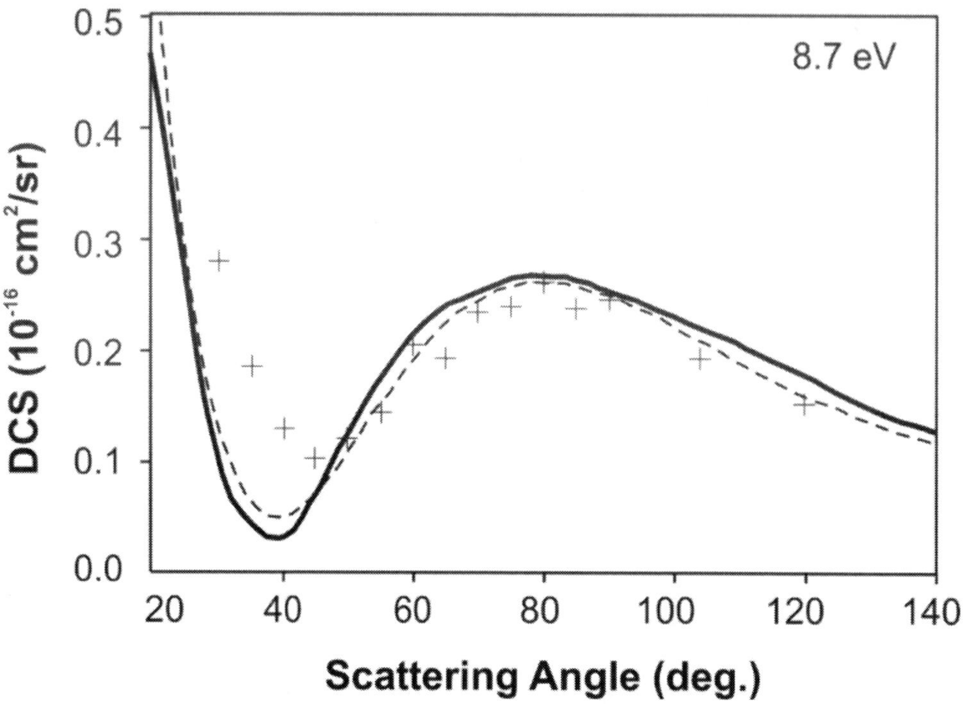

Figure 2. Differential cross sections for e^+ - Ar scattering at 8.7 eV. Calculations: ____, present results; - - -, McEachran and Stauffer [9]. Experiment: +, Kauppila *et al* [17].

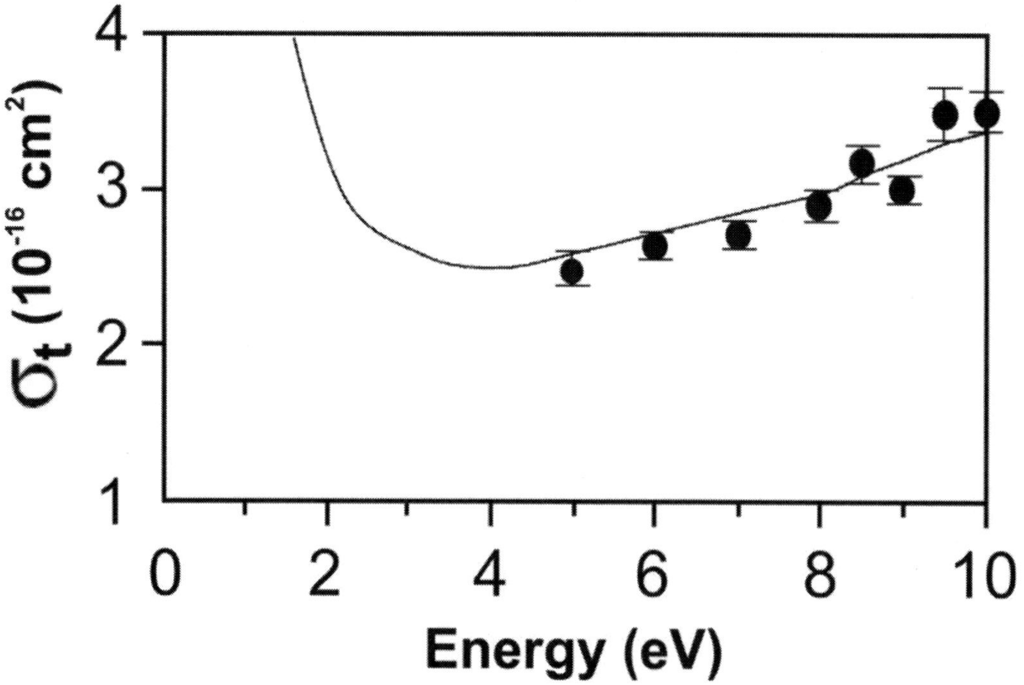

Figure 3. Total cross sections for e^+ - Ar scattering. Calculation: ____, present results. Experiment: ⫮, Coleman *et al* [18].

4. Conclusions

We have presented the results of our calculations of the elastic differential and total cross sections for the scattering of positron with argon atom within a spherical optical potential approach. The calculated DCS reproduces the angular shape and size of the experimental results which is similar to other calculations. The total cross sections are also reported and compared with recent measurements.

Acknowledgement

It is a matter of pleasure to acknowledge the financial support for carrying out this research work from Government of India through the Department of Science and Technology (DST), New Delhi, India under the project no. SR/S2/LOP-0018/2006.

References

[1] Baltzer J C 1994 *Positron Interactions with Atoms, Molecules and Clusters* (Singapore: AC Science)
[2] Kauppila W E and Stien T S 1982 *Can. J. Phys.* **60** 471
[3] Chen S, McEachran R P and Stauffer A D 2008 *J. Phys. B: At. Mol. Opt. Phys.* **41** 025201
[4] Karwasz G P, Brusa R S, Pliszka D, Idziaszek Z and Nowakowska H 2006 *Acta Phys. Pol.* A **110** 605
[5] Salvat F 2003 *Phys. Rev.* A **68** 012708
[6] Mitroy J and Ivanov I A 2002 *Phys. Rev.* A **65** 042705
[7] Reid D D and Wadehra J M 1994 *Phys. Rev.* A **50** 4859
[8] Gianturco F A, Jain A and Rodriguez-Ruiz J A 1993 *Phys. Rev.* A **48** 4321
[9] McEachran R P and Stauffer A D 2009 *J. Phys. B: At. Mol. Opt. Phys.* **42** 075202
[10] Jain A K, Kumar P and Tripathi A N 1994 *Z. Phys.* D **32** 205
[11] Jain A K, Tripathi A N and Jain A 1987 *J. Phys. B: At. Mol. Opt. Phys* **20** L389
[12] O'Connell J K and Lane N F 1983 *Phys. Rev.* A **27** 1893
[13] Millar T M 1993 *CRC Handbook of Chemistry and Physics* vol 74 ed. D R Lide (Boca Raton, FL: CRC press)
[14] McLean A D and McLean R S 1981 *At. Data Nucl. Data Table* **26** 232
[15] Calagero F 1974 *Variable Phase Approach to potential Scattering* (New York: Academic)
[16] Nahar S N and Wadehra J M 1987 *Phys. Rev.* A **35** 2051
[17] Kauppila W E, Kwan C K, Przybyla D, Smith S J and Stein T S 1996 *Can. J. Phys.* **74** 474
[18] Coleman P G, Cheesman N and Lowry E R 2009 *Phys. Rev. Lett.* **102** 173201

An electrostatic slow positron beam apparatus for positron scattering experiments

Kazuaki Nagumo[a,1], **Yuna Nitta**[a], **Daiki Tomida**[b], **Masamitsu Hoshino**[b], **Hiroshi Tanaka**[b], **Yasuyuki Nagashima**[a,2]

[a] *Department of Physics, Tokyo University of Science, 1-3 Kagurazaka, Shinjuku-ku, Tokyo 162-8601, Japan*

[b] *Department of Physics, Sophia University, 7-1 Kioicho, Chiyoda-ku, Tokyo 102-8554, Japan*

E-mail: [1] `j1208652@ed.kagu.tus.ac.jp`, [2] `ynaga@rs.kagu.tus.ac.jp`

Abstract. An electrostatic slow positron beam system has been developed for positron scattering studies. The apparatus uses a remoderator in reflection geometry in conjunction with an electrostatic hemispherical energy analyzer for the brightness enhancement. It is possible to determine absolute total scattering cross section values under magnetic field-free conditions down to a few mG. Measurement of the energy distribution of the positrons that are reemitted from the remoderator and total cross sections for positron-He scattering are presented.

1. Introduction

The scattering phenomena between positrons (e^+) and gas molecules are similar to those between electrons (e^-) and gas molecules at sufficiently high energies. At these high energies the cross sections have the same values for $e^+$and e^-scattering. There are, however, interesting differences at lower energies. Whilst the polarization potential is attractive for these two particles, positrons are repelled by the mean static atomic field. Consequently, the cross sections for e^+scattering are considerably smaller than those for e^-scattering. In addition, the positron is distinguishable from the electron in the target atom or molecule, and therefore exchange effects with the projectile are absent. Furthermore, processes of interest unique to positron scattering include the formation of positronium, the bound state of a $e^+$and an e^-.

The slow positron beams currently used in scattering experiments have far lower intensities than those for electron beams. Since the 1970s, several groups have developed magnetically confined or electrostatic slow positron beams for the measurement of scattering cross sections [1]–[17]. Hybrid beams have also been developed [14]. Though positron elastic differential and ionization cross sections have been measured using electrostatic beams [9]–[13][17], most of the total cross sections have been determined under the influence of external axial magnetic fields. They have been corrected for the forward scattered positrons in the field which cannot be distinguished from unscattered positrons using theoretical angular elastic differential scattering cross sections.

In the present work, we have developed an electrostatic slow positron beam apparatus [18], where slow positrons were transported by an electrostatic lens system. In order to avoid the effect of stray magnetic fields, the apparatus was covered with μ-metal shielding. Therefore, it was possible to determine the cross sections under magnetic field-free conditions down to a few mG. In addition, a remoderator was used in the reflection geometry for brightness enhancement in order to produce a high quality beam with a small diameter and small angular divergence even at low energies. Most of the forward scattered positrons can be distinguished from the unscattered positrons and the cross sections are expected to be obtained without corrections, or with small corrections.

2. Apparatus and experimental approach

2.1. An electrostatic slow positron beam apparatus

A schematic diagram of the electrostatic slow positron beam apparatus is shown in Fig.1.

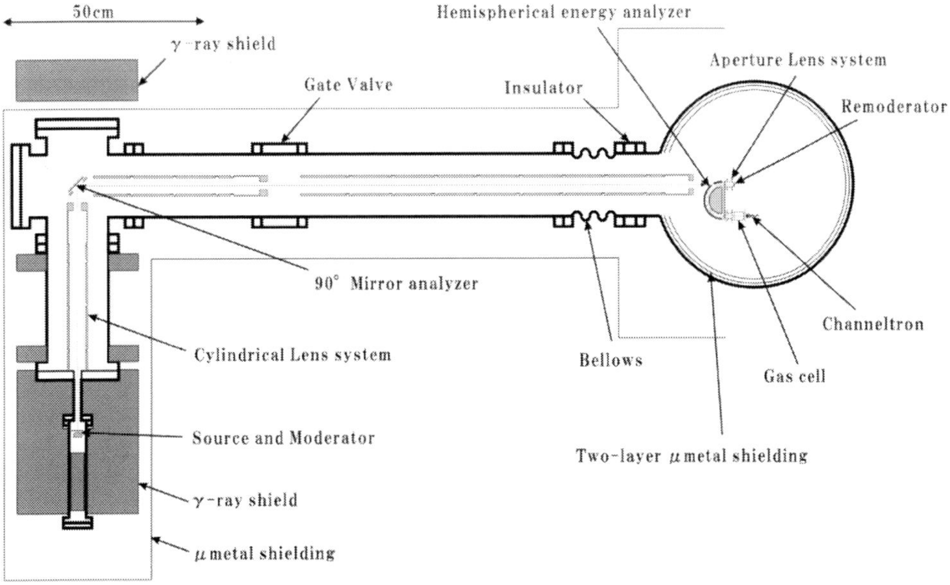

Figure 1. Schematic diagram of the electrostatic slow positron beam apparatus.

Positrons from a 360 MBq ^{22}Na source capsule were moderated with an arrangement of 10 overlapping tungsten meshes. The wires of the meshes were thinned to about 10 μm in diameter by means of an electro-polishing treatment procedure [19]. The moderator meshes were annealed at 2100°C for a few minutes in a separate chamber by passing an electric current through a 25 μm tungsten foil, which encased them in order to remove bulk defects that may act as positron traps. The slow positrons emitted from the moderator were transported by several electrostatic lens elements constructed from aluminum alloy material. The beam was deflected using a 90° mirror analyzer, which was employed to separate the slow positrons from the high-energy positrons and γ-rays. The beam was then transported using further lens elements and focused onto a remoderator with an energy of 4 keV and a diameter of approximately 2-3 mm through an aperture in the electrostatic hemispherical energy analyzer. The remoderator was a tungsten single crystal foil of thickness 2 μm that had been previously annealed at 2100°C in the same chamber used to treat the mesh moderator. Positrons reemitted from the remoderator were deflected and energy selected using the hemispherical energy analyzer [20]. Backscattered positrons from the remoderator were eliminated in this procedure. The beam was transported to a 30.5 mm long gas cell, which had entrance and exit apertures 2 mm in diameter. Positrons that emerged from the cell were detected by a Channeltron.

All parts of the vacuum chamber were made from stainless steel. Three turbo molecular pumps were used with pumping speeds of 250 ℓ/s for the electrostatic transport region, 400 ℓ/s and 250 ℓ/s for the scattering chamber. In order to prevent the effect of stray magnetic fields, the positron gun and the beam transport system were covered with μ-metal shielding. The remoderator, the electrostatic hemispherical energy analyzer, the cell and the Channeltron were also covered with a two-layer μ-metal shielding in the scattering chamber.

2.2. Measurements of the energy distribution of the reemitted positron

When positrons of several keV energy are incident on a single-crystal moderator, they rapidly lose their energies and approach thermal equilibrium in the lattice. In this work, the kinetic energy distribution of the positrons that diffused back to the surface and were reemitted were measured using the electrostatic hemispherical energy analyzer with an analyzer resolution of 40 meV. The results are shown in Fig.2. We determined the value of the positron work function for the tungsten single crystal remoderator to be -2.95 eV. The full-width at half-maximum(FWHM) of the peak was found to be 250 meV.

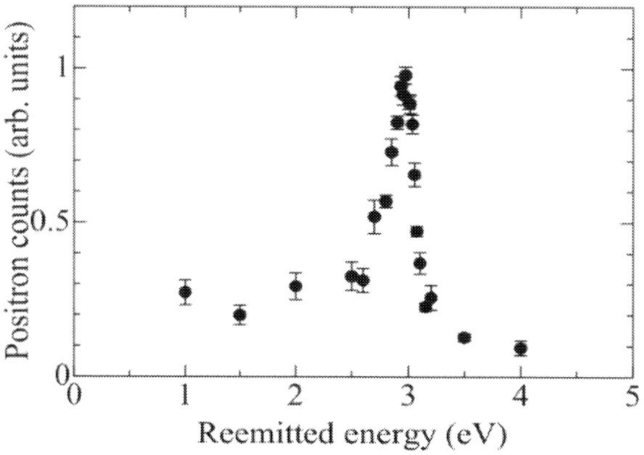

Figure 2. The energy distribution of the reemitted positrons.

2.3. Measurements of total cross sections for positron-He scattering

Absolute total cross sections for positron-He scattering have been measured using a linear transmission method. The energy resolution of the hemispherical analyzer was set at 2 eV in order to increase the count rates. The total resolution for the cross section measurements was 250 meV. The target gas was admitted into the cell via a manually controlled leak valve. The pressure at the center of the cell was determined using an absolute capacitance manometer and typical pressures ranged from 5.0 to 10.0 Pa. Preliminary cross section results are shown in Fig.3, in comparison with some previous measurements[7][16][21][22]. The present results are higher than other measurements in the energy region between 5 and 100 eV. The reason for this discrepancy is not known. Further investigation is necessary.

3. Conclusion

We have developed a magnetic field-free slow positron beam apparatus for the measurement of total scattering cross sections. The energy distribution of the slow positrons for the scattering experiments was found to be approximately 250 meV. Absolute total cross sections for e^+-He in the energy range up to 100 eV have been measured. The preliminary results obtained are higher than other previous measurements.

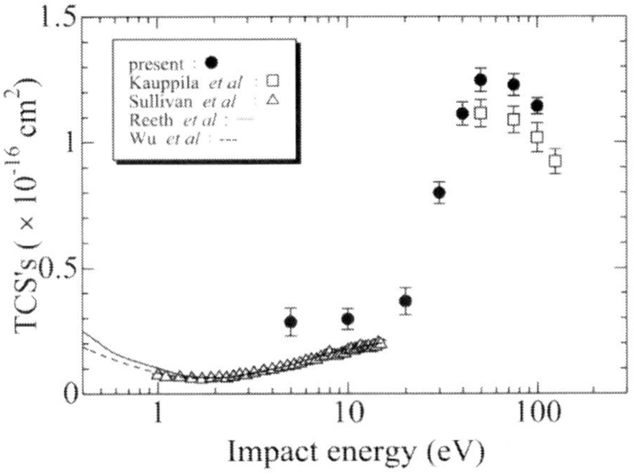

Figure 3. Total cross sections for e^+-He scattering.

4. Acknowledgments

The authors would like to thank Mr. W. Suda and Mr. T. Hakodate for their technical support and encouragement, Prof. S. Buckman, Dr. J. Sullivan and Dr. C. Makochekanwa and Prof. G. Laricchia for valuable discussions. The authors would also like to thank Prof. F. Saito and Prof. T. Hyodo for providing the tungsten mesh moderators and the Sophia University Techno Center Staff for their help with the engineering of our apparatus. This work has been supported by Matsuo Foundation.

5. References

[1] Charlton M and Humberston J 2001 *Positron Physics* (Cambridge: Cambridge University Press)
[2] Costello D G, Groce D E, Herring D F and McGowan J W 1972 *Can. J. Phys.* **50** 23
[3] Canter K F, Coleman P G, Griffith T C and Heyland G R 1973 *J. Phys. B: At. Mol. Phys.* **6** L201
[4] Jaduszliwer B and Paul D A L 1974 *Can. J. Phys.* **3** 249
[5] Brenton A G, Dutton J, Harris F M, Jones R A and Lewis D M 1977 *J. Phys. B: At. Mol. Phys.* **10** 2699
[6] Stein T S, Kauppila W E, Pol V, Smart J H, and Jesion G 1978 *Phys. Rev.* A **17** 1600
[7] Kauppila W E, Stein T S, Smart J H, Dababneh M S, Ho Y K, Downing J P and Pol V 1981 *Phys. Rev.* A **24** 725
[8] Mizogawa T, Nakayama Y, Kawaratani T and Tosaki M 1985 *Phys. Rev.* A **31** 2171
[9] Hyder G M A, Dababneh M S, Hsieh Y -F, Kauppila W E, Kwan C K, Mahdavi-Hezaveh M and Stein T S 1986 *Phys. Rev. Lett.* **57** 2252
[10] Bluhme H, Knudsen H, Merrison J P and Paulsen M R 1998 *Phys. Rev. Lett.* **81** 73
[11] Bluhme H, Frandsen N P, Jacobsen F M, Knudsen H, Merrison J, Paludan K, Poulsen M R 1998 *J. Phys.* B **31** 4631
[12] Bluhme H, Knudsen H, Merrison J P and Nielsen K A 1999 *J. Phys.* B **32** 5237
[13] Bluhme H, Frandsen N P, Jacobsen F M, Knudsen H, Merrison J P, Mitchell R, Paludan K and Poulsen M R 1999 *J. Phys.* B **32** 5825
[14] Karwasz G P, Pliszka D, Brusa R S and Perazzolli C 2005 *Acta Physica Polonica* A **107** 666
[15] Surko C, Gribakin G and Buckman S 2005 *J. Phys. B:At. Mol. Opt. Phys.* **38** R57
[16] Sullivan J P, Makochekanwa C, Jones A, Caradonna P and Buckman S J 2008 *J. Phys. B: At. Mol. Opt. Phys.* **41** 081001
[17] Arcidiacono C, Beale J, Pešicć Z D, Kövér A and Laricchia G 2009 *J. Phys.* B **42** 065205
[18] Suda W, Nagumo K, Tomida D, Hoshino M, Nagashima Y and Tanaka H 2008 *Nucl. Instrum. Methods* B **266** 416
[19] Saito F ,Nagashima Y, Wei L, Itoh Y, Goto A and Hyodo T 2002 *Appl. Surf. Sci.* **194** 13
[20] Fischer D A and Lynn K G 1986 *Phys. Rev.* B **33** 4479
[21] Van Reeth P and Humberston J W 1999 *J. Phys. B: At. Mol. Opt. Phys.* **32** 3651
[22] Wu H, Bray I, Fursa D V and Stelbovics A T 2004 *Phys. B: At. Mol. Opt. Phys* **37** L1

Progress Towards a Positron Reaction Microscope

A. I. Williams[1], Á. Kövér[1,2], D. J. Murtagh[1] and G. Laricchia[1]

[1] UCL, Department of Physics & Astronomy, Gower Street, London, WC1E 6BT, UK.
[2] Institute of Nuclear Research of the Hungarian Academy of Sciences (ATOMKI), Debrecen, Hungary

Abstract. The progress toward the construction of a positron reaction microscope is outlined. The design principles of an electrostatic lens system used to focus and transport the positron beam with a $\sim$ 1 mm diameter spot are briefly discussed. Also presented here are the results obtained from the characterisation of the supersonic gas jet assembly, an increased peaking factor of $\sim$ 4 has been observed along with a centerline density of $\sim 10^{12}$ cm^{-3}.

1. Introduction

The integrated cross-sections for direct ionization and positronium formation have been studied extensively for the inert atoms [1, and references therein] and some molecules [2–5]. However, differential measurements (i.e. resolved in scattered energy and angle) are very scarce being constrained to a handful of measurements over very limited ranges [e.g. 6, 7]. This type of study is expected to contribute to the many fields of investigation which involve positrons, including among others, analysis of energetic events in the galactic center [8], and medical applications such as PET scanners [9]. Reaction microscopes (henceforth referred to as RM) have been used for several years to determine the integral and differential cross-sections of ionizing/fragmentation processes between various projectiles (e.g. photons, ions and anti-protons) and target atoms/molecules [e.g. 10]. However as yet, they have not been used in conjunction with positron impact. The RM setup consists of a projectile beam crossed with a supersonic gas jet target inside a recoil ion spectrometer. The reaction products are usually imaged with one or two Position Sensitive Detectors (PSD). The projectile beam and the supersonic gas jet target have well defined velocities so that the initial conditions of the scattering process are well defined. The recoil ion spectrometer guides the reaction products onto the PSD. This allows the determination of the angular and energy distributions of the product particles.

The ability to determine fully differential cross sections for low energy positron impact on atoms, would allow the study of many interesting phenomena. For example electron capture to the continuum (ECC), which is an ionization process in which the emitted electron is captured by the projectile to a low-lying continuum state. This was first observed in collisions using proton and heavy ion beams [11, 12], where cusp-like structures were seen in the differential cross-section when the velocity of the emitted electron was equal to that of the projectile. This phenomenon has also been observed in positron - H_2 collisions [13]. Interestingly, at 50 eV incident e^+ energy, the emitted electron energy distribution was found to be shifted to lower values in comparison with theory by $\sim$ 1.6 eV [14]. The reason for this is still not fully understood.

In the following report, preliminary results are presented for the characterisation of the positron beam and a helium supersonic gas jet.

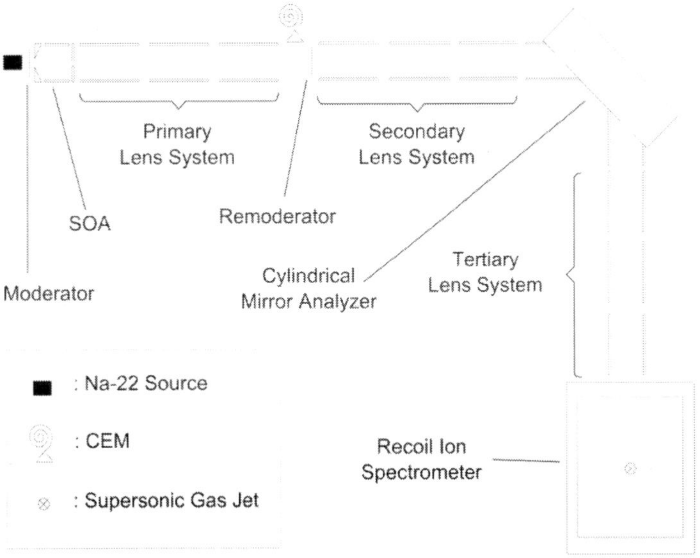

Figure 1. Illustration of the electrostatic lens system designed to create and transport the positron beam

2. Experimental Method & Results

2.1. Beam Transport

The fast positrons emitted from a Na-22 source are impinged onto a tungsten mesh moderator. The re-emitted mono-energetic positrons are transported by an electrostatic field produced by a lens system, which is illustrated in Fig. 1. As shown, the majority of the lenses are made up of three cylindrical elements. This three element design is more commonly known as a zoom lens, as the positions of the object and image can be maintained while the overall acceleration is changed [15]. The beam is focused to a ~ 1 mm spot onto a remoderator (100 nm W foil), from which secondary electrons are also collected to use as a timing signal [16]. A cylindrical mirror analyzer is then used to turn the beam through 90°, which moves the interaction region out of the line of sight of the source and separates out any fast particles emitted from the source. The beam finally intersects the supersonic gas jet at the interaction region.

2.2. Supersonic Gas Jet

Supersonic gas jets are based on the concept of letting a high pressured gas (~ 9 bar) expand through a small nozzle into a low pressure chamber $\sim 10^{-2}$ mbar [17]. The resulting supersonic expanding gas can then be collimated with a skimmer to form a well defined jet, as depicted in Fig.2. Due to the large gas flow rates into such systems several stages of differential pumping are required to keep the pressure in the interaction region low. Such a setup is shown in Fig. 2. Using this type of system, gas jet densities of $\sim 10^{13}$ cm^{-3} have been obtained at the interaction region in previous experiments [18]. In the current setup, it should also be possible to obtain a momentum spread in the transverse direction of $\sim \pm 0.04$ a.u, in comparison with the momentum spread of helium at room temperature which is $\sim \pm 2.7$ a.u.

An important property of a supersonic gas jet is its peaking factor, which is a measure of the angular distribution of the flux from the nozzle. The angular distribution of gas from the nozzle in the current setup can be seen in Fig. 3. The $\cos^8(\theta)$ distribution which fits the data well, relates to a peaking factor of ~ 4 [19]. This is a factor of 2 better than usually obtained [20]. This may be due to the use of a 50 μm electron microscope aperture as the nozzle. It is well known that such apertures have a divergent exit, which may enhance the peaking factor [19].

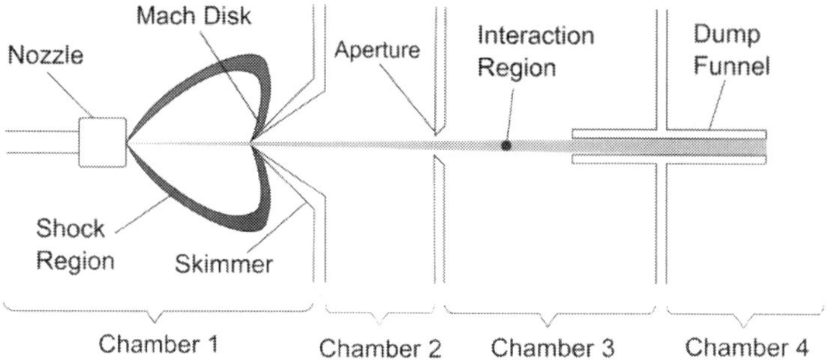

Figure 2. Illustration showing the main features of a supersonic gas jet and the differential pumping stages used in the current setup

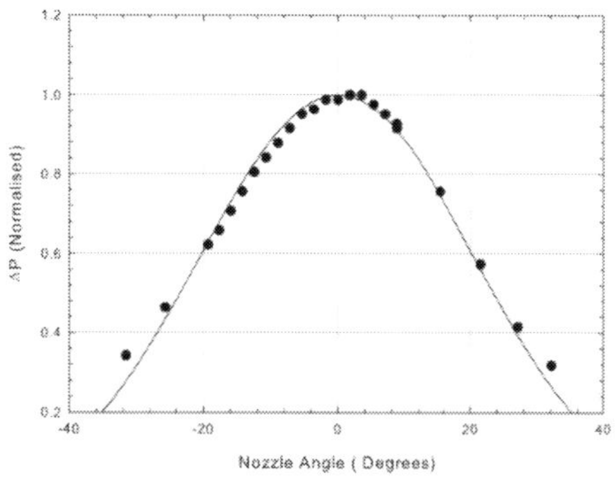

Figure 3. Angular distribution of the supersonic gas jet, ● Experimental results, —— $\cos^8(\theta)$ distribution.

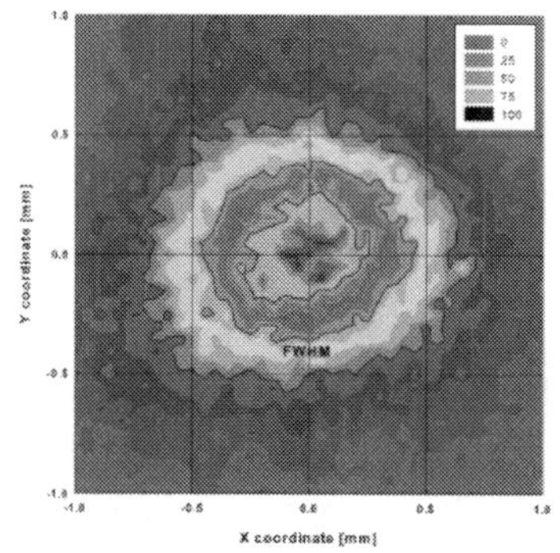

Figure 4. Image of 100 eV positron beam focused on the PSD. The full width at half maximum is ~ 1 mm.

Another important property of the supersonic gas jet is the centreline density, n_c, at the interaction region. This can be calculated from the pressure rise in Chamber 4, ΔP_4, using Eqn.1 [18]:

$$n_c = \frac{\Delta P_4 \cdot S}{v_\infty \pi r^2 k_B T}, \tag{1}$$

where S is the effective pumping speed, $v_\infty = \sqrt{5k_B T/m}$ is the final longitudinal velocity, r is the radius of the jet in the interaction plane and T is the temperature. Initial measurements of the centerline density of the current system have been carried out using this method, with the following initial conditions: 4 bar nozzle pressure; $S \sim 500$ l/s; $r = 6$ mm; $T = 298$ K. The maximum observed pressure rise in chamber 4 was 1.5×10^{-7} mbar which gives $n_c \sim 10^{-12}$ cm^{-3}, about a factor of ~ 5 smaller than calculated from theory.

2.3. Detection

A recoil ion spectrometer will be used to extract the ions from the interaction region onto a RoentDek DLD80 position sensitive detector [21]. Using the DLD80 detector a 100 eV positron beam, produced by the electrostatic lens system (Fig. 2) was imaged and is shown in Fig. 4. The full width at half maximum (FWHM) is ~ 1 mm and ~ 200 remoderated positrons per second have been obtained from a 10 mCi Na-22 source.

3. Conclusion

The progress made towards the production of a positron reaction microscope has been outlined. The first characterisation measurements of the new supersonic gas jet assembly have been presented, along with the first image of a 1 mm positron beam produced with the new electrostatic lens system. Optimisation of the various components is in progress. The aim is to use the equipment outlined above to study fully differential cross-sections for positron-matter interactions. This will include a detailed study of electron capture to the continuum.

Acknowledgments

J.Dumper and R. Jawad for technical assistance, and EPRSC for funding this project under Grant No. EP/E053521/1. A.K. thanks the Hungarian Scientific Research Foundation for support (NKTH-OTKA, Grant No. 67719, OTKA K73703),

References

[1] Laricchia G, Armitage S, Kövér Á and Murtagh D J 2008 *Adv. At. Mol. Opt. Phy.*, *56* (Elsevier) pp 1–47
[2] Kauppila W E, Stein T S, Miller E G and Surdutovich E 2004 *Nucl. Instrum. Methods Phys. Res., Sect. B* **221** 48
[3] Marler J P and Surko C M 2005 *Phys. Rev. A: At. Mol. Opt. Phys.* **72** 062713–10
[4] Murtagh D, Arcidiacono C, Pesic Z and Laricchia G 2006 *Nucl. Instrum. Methods Phys. Res., Sect. B* **247** 92
[5] Cooke D, Murtagh D, Kövér A and Laricchia G 2008 *Nucl. Instrum. Methods Phys. Res., Sect. B* **266** 466
[6] Kövér Á, Laricchia G and Charlton M 1993 *J. Phys. B: At., Mol. Opt. Phys.* **26** L575
[7] Falke T, Brandt T, Kuhl O, Raith W and Weber M 1997 *J. Phys. B: At., Mol. Opt. Phys.* **30** 3247
[8] Belanger G, Goldwurm A, Renaud M, Terrier R, Melia F, Lund N, Paul J, Skinner G and Yusef-Zadeh F 2006 *The Astrophysical Journal* **636** 275
[9] Champion C and Loirec C L 2006 *Phys. Med. Biol.* **51** 1707
[10] Ullrich J, Moshammer R, Dorn A, Dorner R, Schmidt L P H and Schmidt-Bocking H 2003 *Rep. Prog. Phys.* **66** 1463
[11] Rodbro M and Andersen F D 1979 *J. Phys. B: At. Mol. Phys.* **12** 2883
[12] Harrison K and Lucas M 1972 *J. Phys. B: At. Mol. Phys.* **5** L20
[13] Kövér Á and Laricchia G 1998 *Phys. Rev. Lett.* **80** 5309
[14] Kövér Á, Paludan K and Laricchia G 2001 *J. Phys. B: At., Mol. Opt. Phys.* **34** L219
[15] Sise O, Ulu M and Dogan M 2005 *Nucl. Instrum. Methods Phys. Res., Sect. A* **554** 114
[16] Laricchia G, Davies S A, Charlton M and Griffith T C 1988 *J. Phys. E: Sci. Instrum.* **21** 886
[17] Miller R 1988 *Atomic and Molecular Beam Methods. Vol.1* (Oxford University Press)
[18] Götte S, Gopalan A, Bommels J, Ruf M W and Hotop H 2000 *Rev. Sci. Instrum.* **71** 4070
[19] Weaver BD W B 1987 *Rev. Sci. Instrum.* **58** 2115
[20] Beijerinck H C W and Verster N F 1981 *Physica B & C* **111** 327
[21] Spillmann U, Jagutzki O, Spielberger L, Dorner R, Mergel V, Ullmann-Pfleger K and Schmidt-Bocking H 2001 *Physica Scripta* **T92** 225

AUTHOR INDEX

Akimoto, G. 20
Armour, E. A. G. 37
Asai, S. 9, 20
Assafrão, Denise 85
Baker, C. J. 25
Banković, A. 95
Barrachina, R. O. 124
Beddows, D. C. S. 25
Brawley, S. J. 1
Bray, I. 120
Bromley, M. W. J. 68
Brusa, R. S. 110
Buckman, S. J. 95
Campeanu, R. I. 105
Charlton, M. 25, 50
Chiş, V. 105
Cooke, D. A. 31
Cooper, J. N. 37
Danielson, J. R. 73
Darewych, Jurij W. 63
Doser, Michael 56
Dubey, Indira 101
Dujko, S. 95
Fiol, J. 124
Fursa, D. 120
Ghoshal, Arijit 91
Gregory, M. R. 37
Gribakin, G. F. 79
Ho, Y. K. 91
Horbatsch, Marko 63
Hoshino, Masamitsu 134
Idehara, T. 9, 20
Isaac, C. A. 25
Ishida, A. 9, 20
Jain, Arvind Kumar 101, 129
Jonsell, S. 37, 50
Kadyrov, A. S. 120
Kar, Sabyasachi 91
Karwasz, G. P. 110
Kato, K. 20
Kerrigan, S. J. 25
Kobayashi, S. 9, 20
Kobayashi, T. 9, 20
Kövér, Á. 115, 138
Laricchia, G. 1, 31, 115, 138
Malović, G. 95
Marjanović, S. 95
Marler, J. P. 95
Michishio, K. 15
Mitroy, J. 68
Miyamoto, A. 15
Miyazaki, A. 9
Mohallem, José R. 85
Mohan, Harsh 101, 129
Murtagh, D. J. 31, 115, 138
Nagashima, Y. 15, 134
Nagumo, Kazuaki 134
Nagy, L. 105
Namba, T. 9, 20
Nitta, Yuna 134
Ogawa, I. 9, 20
Petrović, Z. Lj 95
Pliszka, D. 110
Plummer, M. 37
Robson, R. E. 95
Sabchevski, S. 9
Saito, H. 9, 20
Sharma, Sunita 129
Suehara, T. 9, 20
Surko, C. M. 73
Šuvakov, M. 95
Tachibana, T. 15
Tanaka, Hiroshi 134
Tanaka, K. 20
Telle, H. H. 25
Terabe, H. 15
Terekidi, Andrei G. 63
Todd, A. C. 37
Tomida, Daiki 134
Tóth, I. 105
Urushizaki, Y. 9
Utamuratov, R. 120
Van Der Werf, D. P. 25, 50
Watkeys, P. R. 25
White, R. D. 95
Williams, A. I. 115, 138
Yamamoto, A. 20
Yoshida, M. 9, 20
Young, J. A. 73

Institute of Physics
Dirac House, Temple Back
Bristol BS1 6BE UK

CURRAN ASSOCIATES INC
proceedings
.com